COSMIC SPACE IS GOD AND PHYSICAL UNIVERSE IS GOD'S DREAM - 3

DR. CHANDRA BHAN GUPTA
B.Sc. (Lko.), M.B.B.S. (Lko.),
M.D. THESIS (MED). (ALLD).,M.R.C.P.(UK),
F.R.C.P. (Edin.), F.R.C.P.(Glasg.),
E.C.F.M.G. CERTIFICATE (U.S.A.)

ISBN: 979-8-6115-1349-1

With Love to My Children

Nirupama

Sujata

Anil

ACKNOWLEDGEMENTS

Let me firstly express my feelings of gratitude to all those who have been instrumental in helping me create this book.

Secondly, I am extremely thankful to my wife Diana, for her help in the preparation of the book despite her various other responsibilities, and to my daughter Sujata for her help. Thirdly, I extend my grateful thanks to my son-in-law, Sunil, whose invaluable technical and practical assistance was proffered notwithstanding his extremely busy schedule.

I would also like to mention many Adwaitic thinkers, scholars, teachers, seekers and students, who after reading my two previous books entitled – **'Adwaita Rahasya : Secrets of Creation Revealed '** and the follow up book entitled – **'Space is The Mind of God : A Scientific Explanation of God and His Abode', 'Cosmic Space is God and Physical Universe is God's Dream,** and **'Cosmic Space is God and Physical Universe is God's Dream - 2',** constantly urged me to write this book.

Finally, I express my sense of intense gratitude to Cosmic Space, who fills the whole cosmos, for choosing me as His instrument to pen this book.

Table of Contents

WHO IS THE CREATOR OF THE PHYSICAL COSMOS ?

The creator of the physical cosmos is a unique consciousness and the name of this unique consciousness is cosmic space, even though, for countless millenniums, embodied human consciousnesses have believed that the creator of the physical cosmos is some mysterious being called god or whatever who resides in some mysterious place called heaven or whatever.

This unique consciousness aka creator of the physical cosmos aka cosmic space is unique in the sense that it is a bodiless consciousness.

By being unique i.e. by being the only one of its kind bodiless consciousness this creator of the physical cosmos aka cosmic space aka god is unlike all the human and animal consciousnesses of the physical cosmos. This is so because all the latter are embodied consciousnesses.

For the embodied consciousnesses of the physical cosmos namely, the human and animal consciousnesses of the physical cosmos, to become bodiless is not possible so long as their physical bodies are alive. However, the moment their physical bodies die, they each become bodiless and immediately merge back into the bodiless consciousness of their source aka cosmic space.

The embodied consciousness of each and every human and animal live-physical-body of the objective cosmos is a driblet, droplet or drop of the bodiless consciousness of their source aka cosmic space aka bodiless consciousness of the creator of the physical cosmos.

These driblets, droplets or drops of consciousness namely, the embodied consciousness of each and every human and animal live-physical-body of the objective cosmos have been provided their passing, fleeting, brief or short-lived individuality, separateness

or separation from the bodiless consciousness of their source aka cosmic space aka bodiless consciousness of the creator of the physical cosmos by the latter itself, that is to say, by their source aka cosmic space aka bodiless consciousness of the creator of the physical cosmos itself and by no one else.

The bodiless consciousness aka cosmic space aka creator of the physical cosmos aka source of each and every driblet, droplet or drop of consciousness which resides inside the live-physical-body of each and every human being and animal of the physical cosmos, has afforded or accorded the gift of individuality (or, if one likes, the gift of separation or separateness from itself) to each and every one of these driblets, droplest or drops of consciousness by encasing each and every one of them into their own or individual physical body, or, if it is preferred, by encasing each and every one of them into their personal or distinctive physical body or, better still, by encasing each and every one of them into their exclusive or unique physical body.

The bodiless consciousness aka cosmic space aka creator of the physical cosmos aka source of each and every driblet, droplet or drop of consciousness which reside inside the live-physical-body of each and every human being and animal, has afforded or granted the gift of individuality or, if one like, has afforded or granted the gift of separation or separateness from itself to these driblets, droplets or drops of its consciousness in order to give them the epithet, moniker or label of human and animal consciousnesses by encasing each and every one of them into their own or individual physical body, or, if it is preferred, by encasing each and every one of them into their personal or distinctive physical body or, better still, by encasing each and every one of them into their exclusive or unique physical body.

What has been said above can be put in another way.

The only one of its kind or unique bodiless consciousness who is the source of all the embodied consciousnesses of the physical cosmos and who also is the creator of the physical cosmos plus who is called cosmic space has itself transiently, temporarily, evanescently or fleetingly severed, separated or cleaved off countless single, separate, independent, individual or discrete driblets, droplets or drops of its own consciousness from itself and encased each and every one of them into their own, individual, personal, distinctive, exclusive or unique physical body in order to create, generate or give rise to variety, diversity, heterogeneity or multiplicity inside its nondescript, featureless, boring or uninteresting plus bodiless consciousness aka cosmic space with a view to amuse, entertain or regale itself and to ward off, keep off, beat off or block its feeling of loneliness and feeling of being unloved.

The task of separating, severing or cleaving off of countless single, separate, independent, individual or discrete droplets, driblets or drops of its own consciousness from itself and encasement of each and every one of them into their own, individual, personal, distinctive, exclusive or unique physical body is achieved by the only one of its kind or unique bodiless consciousness aka cosmic space aka source of all the embodied consciousnesses of the physical cosmos aka creator of the physical cosmos through a very ordinary or common-'o'-garden, consciousnessbal activity on its part, called the activity of daydreaming or oneiricking or, if it is preferred, called the activity of daydreamism or oneirism or, better still, called the activity of consciousnessbal-imagery-making, consciousnessbal-dreamry-making, mental-imagery-making or mental-dreamry-making and nothing else.

Let one now enumerate or catalogue all the other attributes of the only one of its kind or unique bodiless consciousness aka cosmic space aka source of all the embodied consciousnesses of the physical cosmos aka creator of the physical cosmos.

In fact, by enumerating or cataloging all the other attributes of the only one of its kind or unique bodiless consciousness aka cosmic space aka source of all the embodied consciousnesses of the physical cosmos aka creator of the physical cosmos one will be enumerating or cataloging all the other attributes of oneself, that is to say, one will be enumerating or cataloging all the other attributes of one's own embodied consciousness because one is embodied consciousness and embodied consciousness only and nothing but embodied consciousness and embodied consciousness only, notwithstanding or despite one's extraordinary delusion that one is one's physical body and nothing but one's physical body, one's extraordinary delusion which has haunted one from the moment of one's appearance, arrival or emergence in the physical cosmos as a single, separate, independent, individual or discrete embodied consciousness.

The absolute truth is that almost all the attributes which are innate to the only one of its kind or unique bodiless consciousness aka cosmic space aka source of all the embodied consciousnesses of the physical cosmos aka creator of the physical cosmos are also innate to its myriad drops namely, the myriad embodied consciousnesses of the physical cosmos including embodied human consciousnesses.

Having said what one has said above with regards to almost identicalness of all the innate attributes of the only one of its kind or unique bodiless consciousness aka cosmic space aka source of all the embodied consciousnesses of the physical cosmos aka creator of the physical cosmos on one hand, and all the innate attributes of its myriad drops namely, the myriad embodied consciousnesses of the physical cosmos including the embodied human consciousnesses, on the other, one must now clearly state or spell out that there are two vital differences that obtain between the two groups of consciousnesses or which are noteworthy between the two groups of consciousnesses.

These two vital differences must therefore be clearly taken note of by all the embodied human consciousnesses.

These two vital differences between the two groups of consciousnesses or which are noteworthy between the two groups of consciousnesses are as follows :-

Firstly, their respective size i.e. one is a mere drop of consciousness whilst the other is a unique infinite ocean of consciousness.

Secondly, one amongst these two is the only one of its kind or unique bodiless consciousness whilst the other is an ordinary or a common-'o'-garden embodied consciousness.

Let one further elaborate what one means when one says

that there are two vital differences between the only one of its kind or unique bodiless consciousness aka cosmic space aka source of all the embodied consciousnesses of the physical cosmos aka creator of the physical cosmos on one hand and its myriad drops namely, the myriad embodied consciousnesses of the physical cosmos including the embodied human consciousnesses of the physical cosmos, on the other.

One consciousness amongst these two i.e. the only one of its kind or unique bodiless consciousness aka cosmic space aka source of all the embodied consciousnesses of the physical cosmos aka creator of the physical cosmos, is an infinite ocean of consciousness whereas the other i.e. the myriad embodied consciousnesses of the physical cosmos, are mere tiny drops of consciousness which have been severed, cleaved, chopped off or hacked off from the infinite ocean of consciousness aka the only one of its kind or unique bodiless consciousness aka cosmic space aka source of all the embodied consciousnesses of the physical cosmos aka creator of the physical cosmos. Thus, an embodied human consciousness of the physical cosmos is a tiny driblet, droplet or drop of consciousness to whom the following two treatments have been meted out or, who has been subjected to the following two treatments by its source, namely, the infinite ocean of consciousness aka the only one of its kind or unique bodiless consciousness aka cosmic space aka source of all the embodied consciousnesses of the physical cosmos aka creator of the physical cosmos.

Firstly, it has been severed, cleaved, chopped off or hacked off from its source namely, the infinite ocean of consciousness aka the only one of its kind or unique bodiless consciousness aka cosmic space aka the source of all the embodied consciousnesses of the physical cosmos aka the creator of the physical cosmos.

Secondly, it has been encased in a physical body or, if one prefers, secondly, it has been made embodied by this infinite ocean of consciousness aka the only one of its kind or unique bodiless consciousness aka cosmic space aka the source of all the embodied consciousnesses of the physical cosmos aka the creator of the physical cosmos in order to bring into being 6 or 7 billion or whatever individual human consciousnesses in its creation or handiwork called the physical cosmos or, if one prefers, in order to give rise to a mind-boggling or mind-blowing variety, diversity, heterogeneity or multiplicity inside its nondescript, featureless, boring or uninteresting plus bodiless consciousness aka cosmic space with a view to amuse, entertain or regale itself and to ward off, keep off, beat off or block its feeling loneliness and feeling of being unloved.

What has been said above can be put in a slightly

different way.

Thus, an embodied human consciousness of the physical cosmos is a tiny driblet, droplet or drop of consciousness which has been severed, cleaved off, chopped off or hacked off from its source namely, the infinite ocean of consciousness aka the only one of its kind or unique bodiless consciousness aka cosmic space aka the source of all the embodied consciousnesscs of the physical cosmos aka the creator of the physical cosmos, a tiny driblet, droplet or drop of consciousness which has been encased in a physical body or, if one prefers, which has been made embodied by this infinite ocean of consciousness aka the only one of its kind or unique bodiless consciousness aka cosmic space aka the source of all the embodied consciousnesses of the physical cosmos aka the creator of the physical cosmos in order to bring into being 6 or 7 billion or whatever individual human consciousnesses in its creation or handiwork called the physical cosmos or, if one prefers, in order to give rise to a mind-boggling or mind-blowing variety, diversity, heterogeneity or multiplicity inside its nondescript, featureless, boring or uninteresting plus bodiless consciousness aka cosmic space with a view to amuse, entertain or regale itself and to ward off, keep off, beat off or block its feeling loneliness and feeling of being unloved.

The infinite ocean of consciousness aka the only one of its kind or unique bodiless consciousness aka cosmic space aka source of all the embodied consciousnesses of the physical cosmos aka creator of the physical cosmos is eternal as well as non-physical.

The attribute of being eternal on one hand and the attribute of being non-physical on the other, which are possessed by the infinite ocean of consciousness aka the only one of its kind or unique bodiless consciousness aka cosmic space aka the source of all the embodied consciousnesses of the physical cosmos aka the creator of the physical cosmos, are not exclusive to it, personal to it, private to it, especial to it or unique to it and therefore not limited to it.

Instead, the attribute of being eternal on one hand and the attribute of being non-physical on the other, are pan-consciousnessbal attributes or, if it is preferred, are generic attributes which are innate to all the genera of consciousnesses of the physical cosmos or, better still, are such attributes which are universal amongst all the varieties of consciousnesses of the physical cosmos.

Therefore, the attribute of being eternal as well as the attribute of being non-physical are innate or inherent attributes of all the consciousnesses of the physical cosmos and hence, are possessed by all the consciousnesses of the physical cosmos irrespective of whether the consciousness in question is a bodiless consciousness or an embodied consciousness and irrespective of whether the consciousness in question is an infinite ocean of consciousness or a mere tiny drop of this infinite ocean of consciousness.

That is to say, the attribute of being eternal on one hand and the attribute of being non-physical on the other, are possessed by all the consciousnesses of the physical cosmos, irrespective of whether the consciousness in question is the infinite ocean of consciousness aka the only one of its kind or unique bodiless consciousness aka cosmic space aka source of all the embodied consciousnesses of the physical cosmos aka creator of the physical cosmos or, are the embodied human or animal consciousnesses of the physical cosmos which have emanated from this infinite ocean of consciousness aka the only one of its kind or unique bodiless consciousness aka cosmic space aka the source of all the embodied consciousnesses of the physical cosmos aka the creator of the physical cosmos.

In other words, the attribute of being eternal on one hand and the attribute of being non-physical on the other, are possessed by all consciousnesses of the physical cosmos, irrespective of whether the consciousness in question is the infinite ocean of consciousness aka cosmic space or, is a mere tiny drop of consciousness whose source is this infinite ocean of consciousness aka cosmic space.

What has been said above can be put in another way.

The attribute of being eternal, on one hand, and the attribute of being non-physical, on the other, are such attributes which are also possessed by all those consciousnesses which, unlike the unique or the only one of its kind bodiless consciousness aka cosmic space aka the creator of the physical cosmos, are embodied consciousnesses, for example, all embodied human and animal consciousnesses of the physical cosmos.

Thus, the attribute of being eternal on one hand, and the attribute of being non-physical on the other, both of which are possessed by the unique or the only one of its kind bodiless consciousness aka cosmic space aka the creator of the physical cosmos, are also possessed by all the embodied consciousnesses of the physical cosmos namely, all the embodied human and animal consciousnesses of the physical cosmos. There is no exception to this generic or universal rule which is pan-consciousnessbal. That is to say, it applies to all consciousnesses of the physical cosmos. The unique or only one of its kind bodiless eternal and non-physical consciousness aka the infinite ocean of consciousness aka cosmic space aka source of all the embodied consciousnesses of the physical cosmos aka creator of the physical cosmos is innately and therefore eternally or forever dimensionless in nature persona or temper. Being innately and therefore eternally or forever dimensionless in nature persona or temper means that it never needs space for its spatial placement and existence in the manner physical matter of the objective cosmos needs or requires by virtue of being innately dimensional in contour or configuration or, better still, by virtue of being of being innately 3-D or three-dimensional in contour or configuration.

All the embodied eternal and non-physical consciousnesses of the physical cosmos, for example, all the embodied, eternal and non-physical human and animal consciousnesses of the physical cosmos need the pre-presence or pre-existence of space or, absolutely to the point, need the pre-presence or pre-existence of cosmic space, though not for themselves, because they themselves are innately and therefore, eternally or forever dimensionless in nature, persona or temper in the manner of the unique or the only one of its kind, bodiless, eternal and non-physical consciousness aka the infinite ocean of consciousness aka cosmic space aka source of all the embodied consciousnesses of the physical cosmos aka creator of the physical cosmos.

However, the stark or blunt truth is that all the embodied,

eternal and non-physical consciousnesses of the physical cosmos, for example, all the embodied, eternal and non-physical human and animal consciousnesses of the physical cosmos in practice, in reality, in real life or what actually happens as opposed to what is meant or believed to happen, need the pre-presence or pre-existence of space or, absolutely to the point, need the pre-presence or pre-existence of cosmic space. This is the reality owing to the fact that their dimensional or, better still, their 3-D or three-dimensional physical body, casing, sheath or shell needs the pre-presence or pre-existence of space or, absolutely to the point, needs the pre-presence or pre-existence of cosmic space for its spatial placement and existence.

The unique or the only one of its kind, bodiless, eternal and non-physical consciousness aka the infinite ocean of consciousness aka cosmic space aka the source of all the embodied consciousnesses of the physical cosmos aka the creator of the physical cosmos exists in two forms or versions namely, the unexpanded, undistended, undilated or uninflated form or version on one hand, and the expanded, distended, dilated or inflated form or version, on the other.

However, the unique or the only one of its kind, bodiless, eternal and non-physical consciousness aka the infinite ocean of consciousness aka cosmic space aka the source of all the embodied consciousnesses of the physical cosmos aka the creator of the physical cosmos does not exist in both forms or versions at one and the same time. Instead, at any given instant, it exists either in one form or version or, the other, never both.

Presently the unique or the only one of its kind, bodiless, eternal and non-physical consciousness aka the source of all the embodied consciousnesses of physical cosmos aka the creator of physical cosmos exists or is extant, present or existent in its expanded, distended , dilated or inflated form or version.

The expanded, distended , dilated or inflated form or version of the unique or the only one of its kind bodiless, eternal and non-physical consciousness aka the source of all the embodied consciousnesses of physical cosmos aka the creator of physical cosmos has been given the name or tag of cosmic space or, has been give the name or tag of "Brahmandic Aakash" ("Brahmandic Aakash" is the Sanskrit word for cosmic space) by the embodied human consciousnesses of the physical cosmos because of their immense nescience or ignorance of the Adwaitic Truth of Vedantic Cosmology with regards to the true nature of cosmic space or Brahmandic Aakash.

The only one of its kind or unique bodiless eternal and non-physical consciousness aka the creator of the physical cosmos aka the source of all the embodied consciousnesses of physical cosmos has been in existence in its current expanded, distended, dilated or inflated form or version aka cosmic space aka Brahmandic Aakash for some 13.7 billion light years.

That is to say, the only one of its kind or unique bodiless eternal and non-physical consciousness aka the creator of the physical cosmos aka the source of all the embodied consciousnesses of physical cosmos has been in existence in its current expanded, distended, dilated or inflated form or version aka cosmic space aka Brahmandic Aakash from the moment of advent, birth, beginning or dawn of the transient physical cosmos, that is to say, from the moment of advent, birth, beginning or dawn of the transient time which took place 13.7 billion light years ago.

Prior to 13.7 billion light years i.e. prior to the advent, birth, beginning or dawn of the transient physical cosmos or, if one prefers, prior to the advent, birth, beginning or dawn of the transient time (because the advent, birth, beginning or dawn of the transient physical cosmos also heralded or signaled the advent, birth, beginning or dawn of the transient time) was the TIMELESS ERA and during this TIMELESS ERA, the only one of its kind or unique bodiless eternal and non-physical consciousness aka the creator of the physical cosmos aka the source of all the embodied consciousnesses of the physical cosmos was in existence in its unexpanded, undistended, undilated or uninflated form or version.

In the realm of Adwait-Vedanta, the unexpanded, undistended, undilated or uninflated form or version of the only one of its kind or unique bodiless eternal and non-physical consciousness aka the creator of the physical cosmos aka the sourcc of all the embodied consciousnesses of physical cosmos is called the "ADYA" form or version or, if it is preferred, "AADRA" form or version, or, better still, "AADYOT" or "AADYOPAANT" form or version of the only one of its kind or unique bodiless eternal and non-physical consciousness aka the creator of the physical cosmos aka the source of all the embodied consciousnesses of physical cosmos.

That is to say, in the realm of Adwait-Vedanta, the unexpanded, undistended, undilated or uninflated form or version of the only one of its kind or unique bodiless eternal and non-physical consciousness aka the creator of the physical cosmos aka the source of all the embodied consciousnesses of physical cosmos is called its primal, primeval or primordial form or version or, if preferred, is called its original, native or first form or version.

In other words, the unexpanded, undistended, undilated

or uninflated form or version is the "default position" of the only one of its kind or unique bodiless eternal and non-physical consciousness of the creator of the physical cosmos aka the source of all the embodied consciousnesses of physical cosmos.

That is to say, the unexpanded, undistended, undilated or uninflated form or version is the "typical" form or version of the only one of its kind or unique bodiless eternal and non-physical consciousness of the creator of the physical cosmos aka the source of all the embodied consciousnesses of physical cosmos.

The only one of its kind or unique bodiless eternal and non-physical consciousness aka the creator of the physical cosmos aka the source of all the embodied consciousnesses of physical cosmos, exists or occurs or, if it is preferred, is extant or present in its "typical" form or version, that is to say, exists or occurs or, is extant or present in its unexpanded, undistended, undilated or uninflated form or version till it opts or elects to become engaged in its activity of daydreaming or oneiricking or, if it is preferred, till it opts or elects to becomes engaged in its activity of daydreamism or oneirism or, better still, till it opts or elects to becomes engaged in its activity of consciousnessbal-imagery-making, mental-imagery-making, consciousnessbal-dreamry-making or mental-dreamry-making in order to form, forge, construct or create a daydream-stuff-composed, oneiric-stuff-composed, dreamry-stuff-composed, consciousnessbal-imagery-stuff-composed or mental-imagery-stuff-composed cosmos inside the only one of its kind or unique bodiless eternal and non-physical plus innately and therefore, eternally or forever dimensionless consciousness with a view to amuse, entertain or regale itself and to ward off, keep off, beat off or block its feeling of loneliness and the feeling of being unloved.

One absolute fact about the only one of its kind or unique bodiless non-physical and eternal consciousness aka the creator of the physical cosmos aka the source of all the embodied consciousnesses of physical cosmos one must always keep in one's mind. This absolute fact is as follows.

Irrespective of whether the only one of its kind or unique bodiless non-physical and eternal consciousness aka the creator of the physical cosmos aka the source of all the embodied consciousnesses of physical cosmos is in its expanded, distended, dilated or inflated form or version i.e. in the form or version of cosmic space or Brahmandic Aakash or, in its unexpanded, undistended, undilated or uninflated form or version i.e. in its "typical" form or version aka its original, native or first form or version aka its primal, primeval or primordial form or version aka its 'default position', it always is it always will be or, if it is preferred, it eternally is and eternally will be dimensionless and dimensionless only and nothing but dimensionless and dimensionless only, meaning thereby it never does and never will need or require space for its spatial-placement and existence in the manner the dimensional or, better still, the 3-D or three-dimensional physical matter of the objective cosmos needs or requires.

The physical matter of the objective cosmos needs or requires space or, absolutely to the point, needs or requires cosmic space for its spatial placement and existence because it innately is dimensional or, better still, it innately is 3-D or three-dimensional in contour or configuration.

To sum up.

The TIMELESS-ERA-FORM or the TIMELESS-ERA-INCARNATION or the TIMELESS-ERA-SWAROOP or the TIMELESS-ERA-AWATAR of the creator of the physical cosmos is THAT incredible form, incarnation, swaroop or awatar of the creator of the physical cosmos which existed 13.7 billion light years ago. It was the mind-blowing or mind-boggling ERA of the creator of the physical cosmos which existed prior to the advent, birth, beginning or dawn of the current Time and Its Twin, the current Physical Cosmos both of which are transient, temporary, evanescent or ephemeral.

The TIMELESS-ERA-FORM or the TIMELESS-ERA-INCARNATION or the TIMELESS-ERA-SWAROOP or the TIMELESS-ERA-AWATAR of the creator of the physical cosmos is the symbol of the Adwaitic-Core of Vedantic-Cosmology.

The TIMELESS-ERA-FORM or the TIMELESS-ERA-INCARNATION or thc TIMELESS-ERA-SWAROOP or the TIMELESS-ERA-AWATAR of the creator of the physical cosmos represents its dimensionless-quintessence. This dimensionless-quintessence is the quintessence of all consciousnesses irrespective of whether the consciousness in question is the consciousness of the creator of the physical cosmos or that of man.

The dimensionless-quintessence of consciousness is not destroyable.

That is to say, dimensionless-quintessence of consciousness is eternal in the manner consciousness is eternal.

The dimensionless-quintessence of consciousness is not destroyable irrespective of whether the consciousness in question is the consciousness of the creator of the physical cosmos or that of man.

When one says that the dimensionless-quintessence of consciousness is not destroyable one means that this dimensionless-quintessence of consciousness is maintained, preserved or perpetuated eternally irrespective of whether the consciousness is in its expanded, distended, dilated or inflated form or version or in its unexpanded, undistended, undilated or uninflated form or version.

That is to say, the dimensionless-quintessence of consciousness; irrespective of whether the consciousness in question is the consciousness of the creator of the physical cosmos or that of man; is maintained, preserved or perpetuated eternally irrespective of whether the consciousness in question is engaged in its activity of daydreaming or oneiricking or, not engaged in its activity of daydreaming or oneiricking.

In other words, the dimensionless-quintessence of consciousness, irrespective of whether the consciousness in question is the consciousness of the creator of the physical cosmos or that of man, is maintained preserved or perpetuated eternally irrespective of whether the consciousness in question is engaged in its activity of daydreamism, oneirism, consciousnessbal-imagery-making, mental-imagery-making, consciousnessbal-dreamry-making or mental-dreamry-making or not engaged in its activity of daydreamism, oneirism, consciousnessbal-imagery-making, mental-imagery-making, consciousnessbal-dreamry-making or mental-dreamry-making.

TIMELESS-SWAROOP OR TIMELESS-FORM OF THE CREATOR OF THE PHYSICAL COSMOS.

Time and physical matter are conjoint twins.

Additionally, time and physical matter are unique conjoint twins which not only take birth together in a conjoint fashion but also die together in a conjoint fashion because the existence of the two are unyieldingly linked. Only one of them cannot take birth in the cosmos in the absence of the other. For example, physical matter cannot take birth in the absence of its twin called time and vice versa.

What has been said above can be put in another way.

The fate of the cosmic physical matter on one hand and the fate of the cosmic time on the other are unyieldingly interlinked because they innately are an absolutely inseparable conjoint twin. Only one of these conjoint twins cannot have its presence in the cosmos in the absence of thc othcr. That is to say, only one of these conjoint twins cannot have its existence in the non-existence of the other.

The interlinked-fate of the transient time on one hand and the transient physical matter on the other is an inflexible or merciless truth.

Embodied human consciousnesses, (all of whom are eternal in the manner of their source namely cosmic space,) must grasp and constantly ponder over the immutable truth of the interlinked-fate of the transient time on one hand and the transient physical matter on the other till this truth becomes permanently internalized in their consciousness, awareness, sentience, chid, chit, mind, manas, psyche, soul, spirit, atman, self, Aham or "I". This is vital if they are madly keen to decipher the secret of their own eternal existence on one hand and the secret of the eternal existence of cosmic space on the other, plus the secret of the transience or, if it is preferred, plus the secret of the transient existence of cosmic physical matter on one hand and the secret of the transience or the secret of the transient existence of latter's conjoint twin called cosmic time on the other, the transient cosmic physical matter which composes the transient physical bodies of all the embodied, eternal consciousnesses of the cosmos on one hand and the transient physical bodies of all the absolutely unconscious, insentient, lifeless or incapable of feeling or understanding transient moons, planets, stars, black holes, galaxies and the like of the cosmos on the other, all of whom are floating, wafting or levitating plus whirling, twirling or spiralling non-stop in the eternal cosmic space from the moment of their inception or birth in the cosmos and therefore from the moment of inception or birth of the twin of the cosmic physical matter called cosmic time in the cosmos.

All the transient physical items of the cosmos will continue to float, waft or levitate plus whirl, twirl or spiral in the eternal cosmic space till the last moment of their transient existence or presence in the cosmos which will also be the last moment of transient existence or presence of their transient twin called cosmic time.

To sum up.

The inter-linked-status of the existence of the transient

cosmic physical matter in the cosmos on one hand and the existence of the transient cosmic time in the cosmos on the other is unbreakable or inviolable. This truth must be grasped and throughly internalised by all those embodied human, eternal consciousnesses who are madly keen to decipher the secret of their own eternal existence and the secret of the eternal existence of cosmic space on one hand plus the secret of the transient existence of the transient, cosmic physical matter and the secret of the transient existence of the conjoint twin of the latter called the transient cosmic time on the other, the transient, cosmic physical matter which composes the transient physical body of each and every embodied, eternal consciousness of the cosmos and also composes the transient physical body of each and every absolutely unconscious, insentient, lifeless or incapable of feeling or understanding, non-eternal, moons, planets, stars, black holes, galaxies and the like of the cosmos, all of whom are floating, wafting or levitating plus whirling, twirling and spiralling non-stop in the eternal cosmic space and has been doing so from the beginning of the transient cosmic time in the cosmos and will continue to do so till the end of the transient cosmic time in the cosmos.

One must always remember that One as consciousness, awareness, sentience, chid, chit, mind, manas, psyche, soul, spirit, atman, self, Aham or "I" is eternal. It is only the physical body of One which is non-eternal.

One must also remember that the most fundamental and thus the most essential ingredient of the cosmos is the eternal cosmic space which, in turn, is the expanded, distended, dilated or inflated form or version of the unexpanded, undistended, undilated or uninflated form or version of the TIMELESS-SWAROOP or the TIMELESS-FORM of the ETERNAL-CREATOR of the non-eternal cosmic physical matter on one hand and non-eternal cosmic time on the other.

However, it must be pointed out that the transient cosmic time has no objective reality in the manner of the objective reality of the transient cosmic physical matter.

Transient cosmic time is merely a subjective reality.
Nevertheless, the fact of the transient but subjective existence of the transient cosmic time in the transient but objective (or physical) cosmos is felt by the eternal but subjective (or non-physical) human consciousness who is a drop, droplet or driblet of the eternal cosmic space, the eternal cosmic space who is the expanded, distended, dilated or inflated form or version of the unexpanded, undistended, undilated or uninflated form or version of the TIMELESS-SWAROOP or the TIMELESS-FORM of the ETERNAL CREATOR of the transient physical or objective matter of the cosmos.

By the way, eternal cosmic space is also subjective or non-physical i.e., is also immaterial or insubstantial in the manner eternal, subjective, non-physical, immaterial or insubstantial human consciousness is.

MAYA AND ADWAIT VEDANTA - 1.

The Sanskrit word Maya means Magic or Illusion. In the realm of Adwait-Vedanta the words conotes that the material, substantial, physical, concrete or objective world, universe or cosmos of countless phenomena, of countless, moons, planets, stars, black holes, galaxies and the like and of countless embodied consciousness including 6 or 7 billion or whatever embodied human consciousnesses and their sciences, technologies, arts, music, sports and the like, is not as it seems or appears.

The material, substantial, physical, concrete or objective world that one perceives via one's five external sense organs namely eyes, ears, nose, tongue and skin and experiences through one's consciousness, awareness, sentience or mind is misleading as far as its absolutely true nature is concerned. It exists but fundamentally is not what it appears to be to one's five external sense organs and to one's consciousness, awareness, sentience or mind or is not what it is interpreted to be by one's five external sense organs and by one's consciousness, awareness, sentience or mind.

To say that the material universe is Maya or Magic or illusion is not to say that it is unreal or false or it does not exist or it is a lie.

It is to say, instead, that it is not what it seems to be or what it is believed to be by the embodied human consciousnesses.

Maya or Magic or Illusion not only deceives people about the things they think they know. More basically, it limits their knowledge about the things they think they know. Here the word 'things' refers to the material things of the world or, absolutely to the point, refers to the material phenomena of the world.

The word Maya implies or suggests a "magic show", or an illusion where things appear to be present but are not what they seem or appear or, are not what they are believed to be or accepted to be.

From Advait-Vedantic perspective, there are two classes of realities, presences or existences in the world which must be comprehended or fathomed by embodied human consciousnesses or by embodied consciousnesses of Homo sapiens. These are :-

Vyavaharic reality (empirical reality, perceptual reality, phenomenal reality, practical reality, partial reality or relative reality).

And

Parmarthic reality (absolute reality or the extreme end-point reality or the extreme ground-level reality).

Maya, Magic or Illusion is the Vyavaharic reality or the empirical, perceptual, phenomenal, practical, partial or relative reality that entangles embodied human consciousnesses in its web very powerfully.

Maya, Magic or Illusion i.e., Vyavaharic reality or empirical reality entangles embodied human consciousnesses in its web so powerfully indeed that most of them believe and accept from the core of their heart that the Vyavaharic reality or the empirical reality and its enjoyment through the five external sense organs of their physical body and through their consciousness is the sole purpose of their transient embodied existence.

Maya, Magic or Illusion has the power to create a very strong delusion inside the discriminatory faculty of embodied human consciousnesses that the Vyavaharic world or the empirical world is absolutely real, factual or actual or, is a one hundred percent, truly or genuinely existing thing or fact.

As a consequence of this very strong delusion inside the discriminatory faculty of embodied human consciousnesses, a very powerful attachment or better still, a very powerful bondage to the Vyavaharic world or to the empirical world is created inside the discriminatory faculty of embodied human consciousnesses. This prevents the unveiling of the absolutely true, unitary truth — the Cosmic Atman, the Cosmic Spirit or the Cosmic Soul, also known as Brahman, God or the Creator.

The theory of maya was developed in India in the ninth-century by the great Adwaitic thinker and philosopher Adi Shankara.

Maya or Magic or Illusion is a fact in that it is the

appearance of phenomena. Since Brahman, god or the creator is the sole absolute reality or the extreme end-point reality or the extreme ground-level reality, Maya or Magic or Illusion is true in Vyavaharic, empirical, perceptual, phenomenal, practical, partial or relative sense only. However, Maya or Magic or Illusion is not the absolute reality or the extreme end-point reality or the extreme ground-level reality. The absolute reality or the extreme end-point reality or the extreme ground-level reality is the reality forever, while what is Vyavaharic, empirical, perceptual, phenomenal, practical, partial or relative reality is only real for now.

Since Maya or Magic or Illusion is the perceived, observed and experienced, material, substantial, physical, concrete or objective world, it is true in the perceptual and experiential context only but is "untrue" in the absolute or extreme end-point or the extreme Ground-level context of Brahman, God or the Creator.

Maya or Magic or Illusion i.e., Vyavaharic, empirical, perceptual, phenomenal, practical, partial or relative reality aka the perceived, observed and experienced, material, substantial, physical, concrete or objective world is not false, unreal or untrue. On the contrary, it is as real or true as all magics of all magicians are or all illusions of all illusionists are.

However, Maya or Magic or Illusion i.e., Vyavaharic, empirical, perceptual, phenomenal, practical, partial or relative reality aka the perceived, observed and experienced, material, substantial, physical, concrete or objective world clouds the discriminatory faculty of embodied human consciousnesses so much plus the latter namely the embodied human consciousnesses become distracted by its charm so much that they do not crave for the absolute reality, extreme end-point or extreme ground-level reality whose Maya, Magic or Illusion, Vyavaharic, empirical, perceptual, phenomenal, practical, partial or relative reality is; or, whose Maya, Magic or Illusion, the perceived, observed and experienced, material, substantial, physical, concrete or objective world is.

Since the great majority of embodied human consciousnesses do not crave for the absolute reality or, since the great majority of embodied human consciousnesses do not crave for the extreme end-point reality or the extreme ground-level reality who is the Mayakaar or the Magician or the Illusionist whose weaving this Maya or Magic or Illusion called Vyavaharic, empirical, perceptual, phenomenal, practical, partial or relative reality is i.e., whose weaving this Maya or Magic or Illusion called the perceived, observed and experienced, material, substantial, physical, concrete or objective world is, they do not seek it either.

And they find their total fulfillment in this Mayakaar's or magician's or illusionist's Maya or Magic or Illusion only or, if preferred, they find their total fulfillment in this Mayakaar's or magician's or illusionist's created or generated Vyavaharic, empirical, perceptual, phenomenal, practical, partial or relative reality only i.e., in the perceived, observed and experienced, material, substantial, physical, concrete or objective world only.

The absolute truth or the extreme end-point truth or the extreme ground-level truth includes both Maya or Magic or Illusion i.e., Vyavaharic truth (empirical truth, perceptual truth, phenomenal truth, practical truth, partial truth, relative truth or comparative truth) and Parmarthic truth (absolute truth or extreme end-point truth or extreme ground-level truth).

In other words, the absolute truth or the extreme end-point truth or the extreme ground-level truth includes both Maya or Magic or Illusion and the Mayakaar or the magician or the illusionist who has woven this Maya or Magic or Illusion.

With respect to the perceived, observed and experienced, material, substantial, physical, concrete or objective world, the Mayakaar or the magician or the illusionist and its Maya or Magic or Illusion are fundamentally not two but one and one thing only because this Mayakaar's or the magician's or the illusionist's Maya or Magic or Illusion does not exist outside HIM but inside HIM as HIS daydream or oneiric and nothing else. Human beings are HIS daydream or oneiric and nothing else just as is remaining part of the perceived, observed and experienced, material, substantial, physical, concrete or objective world is.

Adwaitins state that the goal of spiritual enlightenment, is to realise the Mayakaar or the magician or the illusionist who has woven this Maya or Magic or Illusion and not to wallow in HIS Maya or Magic or Illusion alone.

That is to say, the goal of spiritual enlightenment, state Adwaitins, is to realise the Brahman, God or the Creator who has created this Maya or Magic or Illusion and not to wallow in HIS Maya or Magic or Illusion alone.

After the end of Maya or Magic or Illusion only the Mayakaar or the magician or the illusionist remains.

MAYA AND ADWAIT VEDANTA - 2

The substantial or concrete cosmos and human experience thereof, is an interplay of the eternal, unchanging ground, aka the supreme or absolute consciousness, aka Brahman or God, on one hand, and, the transient or temporary and constantly changing superstructure, aka the substantial or concrete cosmos, on the other.

The eternal, unchanging ground, aka the supreme or absolute consciousness, aka Brahman or God, manifests itself as the transient or temporary and constantly changing superstructure i.e., the substantial or concrete cosmos and also as, countless eternal but embodied consciousnesses who temporarily live or reside plus play or perform their assigned role inside this transient or temporary and constantly changing superstructure aka the substantial or concrete cosmos as per the will of the eternal, unchanging ground aka the supreme or absolute consciousness, aka Brahman or God.

For example, inside this transient or temporary and constantly changing superstructure aka the substantial or concrete cosmos, 6 or 7 billion or whatever, eternal but embodied human consciousnesses, transiently or temporarily live or reside plus play or perform their assigned role, part or character as per the will of the eternal, unchanging ground aka the supreme or absolute consciousness, aka Brahman or God.

The building material of the transient or temporary and constantly changing superstructure i.e., the building material of the transient or temporary and constantly changing, substantial or concrete cosmos is called the transient or temporary and constantly changing physical or objective matter.

This transient or temporary and constantly changing physical or objective matter, composes the transient or temporary and constantly changing material bodies of all the eternal but embodied consciousness of the transient or temporary and constantly changing superstructure aka the substantial or concrete universe.

For example, this transient or temporary and constantly changing physical or objective matter, composes the transient or temporary and constantly changing material bodies of all the eternal but embodied human consciousnesses who live or reside inside this transient or temporary and constantly changing superstructure as well as play or perform their assigned role, part or character inside this transient or temporary and constantly changing superstructure aka the substantial or concrete cosmos as per the will of the eternal, unchanging ground aka the supreme or absolute consciousness, aka Brahman or God.

The transient or temporary and constantly changing physical or objective matter also composes the transient or temporary and constantly changing material bodies of all the absolutely insentient, insensate, inanimate, unconscious or incapable of feeling or understanding plus lifeless, transient or temporary moons, planets, stars, black holes, galaxies and the like which also live or reside and play or perform all their assigned roles, parts or characters inside this transient or temporary and constantly changing superstructure aka the substantial or concrete cosmos, as per the will of the eternal, unchanging ground aka the supreme or absolute consciousness, aka Brahman or God.

The transient or temporary and constantly changing superstructure aka the substantial or concrete cosmos distracts the eternal but embodied human consciousnesses so much by its allure, appeal, or charm that these eternal but embodied human consciousnesses fail to notice the presence of the eternal, unchanging ground aka the supreme or absolute consciousness, aka Brahman or God amidst them. This eternal, unchanging ground aka the supreme or absolute consciousness aka Brahman or God amidst them is none other than the entity whom these eternal but embodied human consciousnesses address as cosmic space.

What has been said above can be put in another way.

The transient or temporary and constantly changing superstructure aka the substantial or concrete cosmos distracts the eternal but embodied human consciousnesses so much by its allure, appeal or charm that these eternal but embodied human consciousnesses fail to notice that the transient or temporary and constantly changing superstructure aka the substantial or concrete cosmos is floating, wafting or levitating plus whirling, twirling or spiralling non-stop inside the supreme or absolute consciousness aka Brahman or God aka cosmic space as the latter's daydream or oneiric.

The above described nescience, obtuseness or denseness of most of the embodied but eternal, human consciousnesses prevails with regards to the eternal, unchanging ground aka the supreme or absolute consciousness, aka Brahman or God, aka cosmic space despite the fact that this eternal, unchanging ground aka the supreme or absolute consciousness, aka Brahman or God, aka cosmic space lives amidst them and is visible to them every moment of their lives.

In other words, the charm or beauty of the transient or temporary and constantly changing superstructure aka the substantial or concrete cosmos is so immense or overwhelming that it succeeds in overpowering the discriminatory faculty of most of the embodied but eternal human consciousnesses completely. As a result, most of the embodied but eternal human consciousnesses are not able to decipher, decode or work out the real identity of cosmic space, that the latter is the one and only, eternal, unchanging ground aka the supreme or absolute consciousness aka Brahman or God. This happens despite the fact that all the embodied but eternal human consciousnesses see the presence of this cosmic space in their midst every moment of their lives.

To repeat.

The charm or beauty of the transient or temporary and

constantly changing substantial or concrete cosmos hides the true identity of cosmic space aka the eternal, unchanging ground aka the supreme or absolute consciousness aka Brahman or God from the embodied but eternal human consciousnesses even though cosmic space is constantly visible to them all the time and even though the embodied but eternal human consciousnesses realise that without the pre-presence or pre-existence of cosmic space the presence or existence of the transient or temporary and constantly changing superstructure aka the substantial or concrete cosmos will not be achievable, attainable, accomplishable or within the bounds of possibility.

Here one more truth must be clearly understood by all the embodied but eternal human consciousnesses as follows.

The six or seven billion or whatever embodied but eternal human consciousnesses themselves are intrinsically or inherently not different from the eternal, unchanging ground aka cosmic space aka the supreme or absolute consciousness aka Brahman or God because each and every one of them, in fact, is an absolutely pure or pristine section, segment, part or potion of this incredible cosmic space aka the eternal, unchanging ground aka the supreme or absolute consciousness aka Brahman or God; the eternal, unchanging ground aka the supreme or absolute consciousness aka Brahman or God aka cosmic space who is the source or fountainhead or, if it is preferred, who is the creator plus who also is constantly supporting and sustaining this transient or temporary and constantly changing superstructure aka the substantial or concrete cosmos of which embodied but eternal human consciousnesses are a constituent part albeit an extremely important constituent part on account of their immense intelligence which is more than that possessed by any other living being.

The knowledge of the eternal, unchanging ground aka the supreme or absolute consciousness aka Brahman or God by knowing that cosmic space is this eternal, unchanging ground aka the supreme or absolute consciousness aka Brahman or God i.e., the knowledge that the eternal, unchanging ground aka the supreme or absolute consciousness aka Brahman or God is none other than the entity called cosmic space and by knowing that all the embodied but eternal consciousnesses, for example, all the embodied but eternal human consciousnesses are absolutely pure or pristine, section, segment, part or potion of this incredible cosmic space aka the eternal, unchanging ground aka the supreme or absolute consciousness aka Brahman or God, is "true knowledge" or rather, is "complete knowledge".

However, the deep or detailed knowledge of the transient or temporary and constantly changing physical or objective matter which composes the transient or temporary and constantly changing, material bodies of all the eternal but embodied consciousness of the transient or temporary and constantly changing physical cosmos on one hand, and the transient or temporary and constantly changing material bodies of all the absolutely insentient, insensate, inanimate, unconscious or incapable of feeling or understanding plus lifeless, transient or temporary and constantly changing moons, planets, stars, black holes, galaxies and the like, of the transient or temporary and constantly changing physical cosmos on the other, is not the "true knowledge" or rather, is not the "complete knowledge".

The above is the case because the deep or detailed knowledge of the absolutely insentient, insensate, inanimate, unconscious or incapable of feeling or understanding plus lifeless physical or objective matter of the substantial or concrete cosmos is incapable of revealing to embodied but eternal human consciousnesses either the true identity of themselves or the true identity of the eternal, unchanging ground aka the supreme or absolute consciousness aka Brahman or God aka cosmic space.

However, the deep or detailed knowledge of the absolutely insentient, insensate, inanimate, unconscious or incapable of feeling or understanding plus lifeless physical or objective matter of the substantial or concrete cosmos is useful for the empirical, practical, pragmatic or Vyavaharic life of the embodied but eternal human consciousnesses during their temporary sojourn or stopover in the substantial or concrete cosmos but this knowledge will not lay open to them or will not expose to them the complete mystery which stands hiding behind the substantial or concrete cosmos on one hand and behind the eternal, unchanging ground aka the supreme or absolute consciousness aka Brahman or God aka cosmic space on the other, both of which are existent plus visible in the current cosmos.

Since the charm or beauty of the temporary and constantly changing superstructure aka the substantial or concrete cosmos obscures or eclipses the true identity of cosmic space aka the eternal, unchanging ground aka the supreme or absolute consciousness aka Brahman or God from the embodied but eternal human consciousnesses, even though this cosmic space aka the eternal, unchanging ground aka the supreme or absolute consciousness aka Brahman or God is constantly visible to embodied but eternal human consciousnesses every moment of their lives and even though the embodied but eternal human consciousnesses realise or accept that without the pre-presence or pre-existence of cosmic space aka the eternal, unchanging ground aka the supreme or absolute consciousness aka Brahman or God, the birth and continued existence of the charming or beautiful, temporary and constantly changing superstructure aka the substantial or concrete cosmos will not be possible, the embodied but eternal human consciousnesses continue to ignore or rather continue to play down or pretend as if cosmic space were unimportant or, continue to downgrade, demote or lower the importance or status of cosmic space aka the eternal, unchanging ground aka the supreme or absolute consciousness aka Brahman or God as compared to the importance or status they accord to the charming or beautiful, temporary and constantly changing superstructure aka the substantial or concrete cosmos. On account of this, the charming or alluring, temporary and constantly changing superstructure aka the substantial or concrete cosmos is designated or labeled as Maya or Magic or Illusion in the realm or domain of Adwait-Vedanta because the charm or the allurement of the constantly changing superstructure aka the substantial or concrete cosmos obscures or eclipses the true identity of cosmic space from the embodied but eternal human consciousnesses. Not only this, the charm or allurement of the constantly changing superstructure aka the substantial or concrete cosmos obscures or eclipses the true identity of the embodied but eternal human consciousnesses themselves from themselves.

The goal of the embodied but eternal human

consciousnesses should be to realise that the pleasing or endearing, temporary and constantly changing superstructure aka the substantial or concrete cosmos is merely an extremely attractive, appealing, beguiling or seducing enchantress, temptress, seductress, sorceress, siren or spellcaster or, is merely an extremely enticing, tempting or alluring Maya or Magic or Illusion and therefore should be accepted by them as such or should be accepted by them essentially or fundamentally in this form or manner only. Consequently, despite it being extremely inviting, captivating, glamorous or seductive, it should be indulged-in, revelled-in, enjoyed or appreciated by the embodied but eternal human consciousnesses in only very measured or calculated doses and must not be taken by them as being the "be all and end all" of their lives or, as being the "alpha and omega" of their lives or, as being the "beginning and end" of their lives.

During their unsought, unrequested or unsolicited, transient or temporary, residence, stay, sojourn or stopover inside the extremely enchanting, beguiling, charming, fascinating, intriguing, or tantalising, substantial or concrete cosmos aka the transient or temporary and constantly changing superstructure, the embodied but eternal human consciousnesses must realise the truth of their own immortality on one hand, and the truth of the real identity of cosmic space on the other. The truth of the real identity of cosmic space is that the latter is the eternal, unchanging ground of the extremely enchanting, beguiling, charming, fascinating, intriguing, or tantalising, temporary and constantly changing superstructure aka the substantial or concrete cosmos. In other words, cosmic space is Brahman or God i.e., is the supreme or absolute consciousness who is the creator, maker or progenitor or, if it is preferred, who is the source or fountainhead plus the support, sustainer or maintainer of the extremely enchanting, beguiling, charming, fascinating, intriguing, or tantalising, temporary and constantly changing superstructure aka the substantial or concrete cosmos.

To repeat.

During their transient sojourn or stopover in the substantial or concrete cosmos, embodied but eternal human consciousnesses must realise the truth of their own immortality on one hand and the truth of the real identity of cosmic space on the other. The truth of the real identity of cosmic space is that it is the eternal, unchanging ground of the temporary and constantly changing superstructure aka the objective or material cosmos. In other words, cosmic space is Brahman or God i.e., is the supreme or absolute consciousness who is the source, support and sustainer of the temporary and constantly changing superstructure aka the objective or

material cosmos

MAYA AND ADWAIT VEDANTA - 3.

The term "Complete Knowledge, Insight, Wisdom, Comprehension or Enlightenment" must include Vyavaharic knowledge i.e., Empirical, Practical, Pragmatic, Physical, Material or Objective knowledge on one hand and Parmarthic knowledge i.e., Consciousnessbal, Awarenessbal, Sentiencel, Brahmanic or Godly knowledge on the other.

The term "Complete Knowledge, Insight, Wisdom, Comprehension or Enlightenment" necessarily includes the understanding of the eternal, unchanging ground i.e. necessarily includes the understanding of consciousness, awareness, sentience, Brahman, God, Absolute or Supreme upon which the temporary and constantly changing superstructure i.e., material, substantial, physical, objective or concrete cosmos is standing or existing or, is erect or upright, or, inside which the material, substantial, physical, objective or concrete cosmos has its foundation or root.

The epithet, sobriquet, nickname or tag of "Maya" or "Magic" or "Illusion" is applied to the material, substantial, physical, objective or concrete cosmos in the realm or domain of Adwait-Vedanta.

The epithet, sobriquet, nickname or tag of "Maya" or "Magic" or "Illusion" is applied to the material, substantial, physical, objective or concrete cosmos because the latter is not as it seems.

The material, substantial, physical, objective or concrete cosmos that one experiences is misleading as far as its true nature is concerned.

The material, substantial, physical, objective or concrete cosmos, is both real and unreal because it exists but is not what it appears to be.

To say that the material, substantial, physical, objective or concrete cosmos is "Maya" or "Magic" or "Illusion" is not to say that it is unreal.

Instead it is that the material, substantial, physical, objective or concrete cosmos is not what it is taken to be.

Or that, it is something that is being constantly made and unmade.

Or that, it is something which is being constantly revised, remade, reshaped, redesigned, restyled, revamped, reworked, refashioned, remodelled, reordered, reconstructed or reorganised.

Or that, it is something which is being constantly recycled, reprocessed or recirculated.

Or that, it is something which is being constantly transformed, transfigured or transmuted.

Or that, which is constantly changing.

Or that, which is a chimera, fantasy, dream or figment of imagination of cosmic space aka Brahman or God, aka Ultimate or Supreme consciousness.

Or that, which is not what it is drummed up to be.

Or that, which is not what it is made up, cooked up, vamped up, or trumped to be.

Or that, which is not what one has been hoaxed or conned into believing it is.

Or that, which is not what one has been hoodwinked or bamboozled into believing it is.

Or that, which is not what one has been fooled into believing it is.

Or that, which is not what one has been duped or deceived into believing it is.

"Maya", "Magic" or "Illusion" not only deceives people about things they think they know; more basically, it limits their knowledge about things they think they know.

More to the point, Maya", "Magic" or "Illusion" deceives people about things which are referred to as being material, substantial, physical, objective or concrete in nature.

Furthermore, "Maya", "Magic" or "Illusion" deceives

people or, rather, limits the knowledge of people about the "Mayakaar", "Magician" or "Illusionist" who is behind this Maya", "Magic" or "Illusion" or, who is behind those things which are referred to as being material, substantial, physical, objective or concrete in nature.

That is to say, Maya", "Magic" or "Illusion" also deceives people or, rather, limits the knowledge of people vis-a-vis the "Mayakaar", "Magician" or "Illusionist" who is behind this Maya", "Magic" or "Illusion" or, absolutely to the point, who is behind the formation, creation, generation or genesis of this Maya", "Magic" or "Illusion" aka material, substantial, physical, objective or concrete cosmos.

What has been said above can be put in another way.

Maya", "Magic" or "Illusion" limits the knowledge of people with regards to the "Mayakaar", "Magician" or "Illusionist" who is the source or fountainhead of this Maya", "Magic" or "Illusion" aka material, substantial, physical, objective or concrete cosmos.

"Maya", "Magic" or "Illusion" i.e., material, substantial, physical, objective or concrete cosmos always exists inside cosmic space aka Brahman or God aka the Ultimate or Supreme Consciousness & never outside it because it is latter's daydream or oneiric.

Sometimes this Maya", "Magic" or "Illusion" i.e., sometimes this material, substantial, physical, objective or concrete cosmos exits inside cosmic space aka Brahman or God aka the Ultimate or Supreme Consciousness in patent, evident, obvious, conspicuous, noticeable, visible, manifest, observable or perceptible form as is presently the case.

But at other times, this Maya", "Magic" or "Illusion", i.e., this material, substantial, physical, objective or concrete cosmos exits inside cosmic space aka Brahman or God aka Ultimate or Supreme Consciousness in latent, dormant, quiescent, hidden, concealed, unrevealed, unexpressed, indiscernible, invisible, unseen, veiled, masked, potential or imperceptible form.

At the present moment this Maya", "Magic" or "Illusion", i.e., this material, substantial, physical, objective or concrete cosmos is a perceived reality inside cosmic space aka Brahman or God aka Ultimate or Supreme consciousness but at any instant it can become a concealed reality inside this cosmic space aka Brahman or God aka Ultimate or Supreme consciousness as per the whim, fancy or mood of this cosmic space aka Brahman or God aka Ultimate or Supreme consciousness.

By its powerful charm, beauty, glamour, appeal, allure, desirability, seductiveness, magnetism, charisma or enticement; this Maya", "Magic" or "Illusion", i.e., this material, substantial, physical, objective or concrete cosmos does not allow the embodied human consciousnesses to discern, glimpse, notice, observe, perceive, spot, see, survey, scan, view, watch, witness, make out, take note of, pay attention to or pay heed to the eternal, unchanging ground aka cosmic space aka Brahman or God aka the Absolute or Supreme consciousness who is the eternal reality upon which the temporary, constantly changing superstructure i.e., the material, substantial, physical, objective or concrete cosmos aka Maya", "Magic" or "Illusion" is standing or existing or, is erect or upright, or, inside which the material, substantial, physical, objective or concrete cosmos aka Maya", "Magic" or "Illusion" has its foundation or root.

Maya", "Magic" or "Illusion" i.e., physical, objective or concrete matter which composes the physical, objective or concrete cosmos and which is much loved and admired by the embodied human consciousnesses is absolutely insentient, insensate, inanimate, unconscious or incapable of feeling or understanding plus lifeless thing.

In stark contrast, totally ignored or taken no notice of, payed no attention to, payed no heed to, passed over, glossed over, brushed aside, overlooked or disregarded eternal, unchanging ground aka cosmic space aka Brahman or God aka Absolute or Supreme consciousness is supremely conscious, aware, sentient or sensate being.

Maya", "Magic Show" or "Illusion" i.e., material, substantial, physical, objective or concrete cosmos is a "created" entity whereas the eternal, unchanging ground aka cosmic space aka Brahman or God aka Absolute or Supreme consciousness is the "creator" of this "created" entity namely the material, substantial, physical, objective or concrete cosmos.

Much loved and admired Maya", "Magic Show" or "Illusion" i.e., material, substantial, physical, objective or concrete cosmos is the "effect".

In contrast, totally ignored or taken no notice of, payed no attention to, payed no heed to, passed over, glossed over, brushed aside, overlooked or disregarded, eternal, unchanging ground aka cosmic space aka Brahman or God aka Absolute or Supreme consciousness is the "cause" of this much loved and admired "effect" aka Maya", "Magic Show" or "Illusion" aka material, substantial, physical, objective or concrete cosmos.

Totally ignored or taken no notice of, payed no attention to, payed no heed to, passed over, glossed over, brushed aside, overlooked or disregarded, eternal, unchanging ground aka cosmic space aka Brahman or God aka Absolute or Supreme consciousness is the fundamental, central or elemental truth.

In contrast, much loved and admired Maya", "Magic Show" or "Illusion" i.e., material, substantial, physical, objective or concrete cosmos is the auxiliary, accessory, secondary, trivial or nonessential truth whose source or fountainhead is the totally ignored or taken no notice of, payed no attention to, payed no heed to, passed over, glossed over, brushed aside, overlooked or disregarded, eternal, unchanging ground aka cosmic space aka Brahman or God aka Absolute or Supreme consciousness aka the fundamental, central or elemental truth.

Much loved and admired Maya", "Magic Show" or "Illusion" i.e., material, substantial, physical, objective or concrete cosmos is born, changes, evolves, dies with time, from circumstances or due to invisible principles of nature. The source or fountainhead of these invisible principles of nature is the totally ignored or taken no notice of, payed no attention to, payed no heed to, passed over, glossed over, brushed aside, overlooked or disregarded, eternal, unchanging ground aka cosmic space aka Brahman or God aka Absolute or Supreme consciousness aka the fundamental, central or elemental truth.

Cosmic space aka Brahman or God aka totally ignored or taken no notice of, payed no attention to, payed no heed to, passed over, glossed over, brushed aside, overlooked or disregarded, eternal, unchanging ground is timeless, unaffected, absolute and resplendent consciousness.

Much loved and admired "Maya", "Magic Show" or "Illusion" i.e., material, substantial, physical, objective or concrete cosmos is the possibility, pre-existing within cosmic space aka Brahman or God aka totally ignored or taken no notice of, payed no attention to, payed no heed to, passed over, glossed over, brushed aside, overlooked or disregarded, eternal, unchanging ground, just like the possibility of a future tree pre-exists in the seed of the tree.

Cosmic space aka Brahman or God aka totally ignored or taken no notice of, payed no attention to, payed no heed to, passed over, glossed over, brushed aside, overlooked or disregarded, eternal, unchanging ground, is not a concealed, camouflaged, hidden, invisible, shrouded, unseen or veiled reality. On the contrary, it is a seeable reality which is seen all the time by all embodied human consciousnesses via the medium of their physical eyes, of but due to the iron grip or hold of "Maya", "Magic Show" or "Illusion" i.e., material world, it is never given any attention.

Nature or the material world is Maya, Magic or illusion. Brahman or God aka the unnoticed, eternal, unchanging ground aka cosmic space, is the Mayakar or the Magician or the Illusionist.

Embodied human consciousnesses are infatuated with "Maya", "Magic Show" or "illusion" i.e., material world aka nature and thus they create for themselves bondage to this "Maya", "Magic Show" or "illusion" i.e., material world aka nature. This drastic mistake on their part is all due to their delusion that this "Maya", "Magic Show" or "illusion" i.e., material world aka nature is actual, factual, real, genuine or authentic.

For freedom and liberation from anxiety, from uncertainty as to what will happen to one after the death of one's material body, one must seek true insight or true knowledge of the eternal, unchanging ground i.e. the supreme consciousness aka cosmic space aka Brahman or God upon which the temporary, constantly changing superstructure i.e., material world or magic or maya is standing.

When a rope is not perceived distinctly in the dark, it is erroneously imagined as a snake. In the same manner consciousness is erroneously imagined as the material universe.

When the rope i.e., consciousness is distinctly perceived, and the erroneous imagination i.e., material world, is withdrawn, only the rope i.e., consciousness remains.

When consciousness is distinctly perceived, then the physical cosmos is understood as mere Maya or Magic which Brahman aka cosmic space i.e., the eternal, unchanging ground has created inside itself through the instrumentality of daydreaming or oneiricking on its part in order simply to amuse, entertain or regale itself, nothing more, nothing less.

PHYSICAL MATTER IS A DAY-DREAMAL OR ONEIRICAL STUFF & NOT AN ACTUAL STUFF.

On account of their nescience or ignorance of the eternal cosmological truth, when embodied human consciousnesses of planet earth assert that physical matter of cosmos is an actual stuff and follow this assertion by insisting that cosmos is made of this actual stuff namely physical matter which is also known as material, substantial, objective or concrete stuff, they absolutely overlook the eternal cosmological truth that there is nothing in the cosmos which is objective, concrete, material, substantial or physical in nature. On the contrary, every item of the cosmos is subjective, day-dreamal, oneirical, consciousnessbal or awarenessbal in nature only and nothing else.

Hence, what embodied human consciousnesses of planet earth stereotype or brand as being physical, objective, or, concrete matter, thing or stuff, the absolute or, the extreme-ground-level or the extreme-end-point truth is that it is not at all physical, objective or concrete matter, thing or stuff.

Instead, it is a day-dreamal or oneirical construct and nothing but a day-dreamal or oneirical construct.

That is to say, it is a subjective construct only and nothing else.

To re-phrase. It is a day-dreamal or oneirical truth only and nothing else.

What is labeled or designated as being the physical matter by embodied human consciousnesses of planet earth, is not at all an actual construct i.e. it is not at all a real, material, substantial or concrete construct because, as per eternal cosmological truth, it is made of nothing else but day-dream stuff or oneiric stuff of cosmic space aka God, Brahman, Creator or Maker of cosmos.

What has been said above can be put in another way.

The entire cosmos is made of consciousnessbal stuff or awarenessbal stuff of cosmic space aka God, Brahman, Creator or Maker of cosmos albeit consciousnessbal or awarenessbal stuff of cosmic space aka God, Brahman, Creator or Maker of cosmos in condensed, compressed, compacted or congealed form, that is to say, consciousnessbal or awarenessbal stuff of cosmic space aka God, Brahman, Creator or Maker of cosmos which has been condensed, compressed, compacted or congealed by cosmic space aka God, Brahman, Creator or Maker of cosmos through the act of daydreaming or oneiricking on its part.

Hence, what is labeled or designated as "material", "substantial", "physical", "objective", or "concrete" matter, thing, or stuff by embodied human consciousnesses of planet earth, it is nothing of the sort.

Instead, it is a condensed, compressed, compacted or congealed form or version of a section, segment, part, or portion of consciousness of cosmic space aka God, Brahman, Creator or Maker of cosmos, a section, segment, part, or portion of consciousness of cosmic space aka God, Brahman, Creator, or Maker of the cosmos which has been condensed, compressed, compacted or congealed by it through the act of daydreaming or oneiricking on its part.

Embodied human consciousnesses of planet earth, stereotype or brand day-dreamal or oneirical stuff or, consciousnessbal or awarenessbal stuff of cosmic space aka God, Brahman, Creator, or Maker of the cosmos as being one hundred percent, actual, factual, real, authentic or genuine stuff or, if it is preferred, being one hundred percent, material, substantial, physical, objective or concrete stuff on account of their nescience or ignorance of the eternal cosmological truth or, on account of their nescience or ignorance of the extreme-ground-level cosmology or, the extreme-end-point cosmology which is in motion non-stop in the cosmos, unseen or unnoticed by any human eyes.

This extreme-ground-level cosmology or, this extreme-end-point cosmology is in motion non-stop for the purpose of creation of the cosmos on one hand and for the purpose of its constant modification on the other as per fancy, mood or whim of cosmic space aka God, Brahman, Creator or Maker of the cosmos.

The eternal cosmological truth or, the nitty-gritty, nub or

quintessence of extreme-ground-level or extreme-end-point of cosmology is the following.

What is stereotyped or branded as being one hundred percent actual, factual, real, genuine or authentic physical matter or physical substance or, if it is preferred, what is stereotyped or branded as being one hundred percent actual, factual, real, genuine or authentic, objective or concrete matter or, objective or concrete substance, in truth, is nothing of the sort.

Instead, it is an absolutely day-dreamal or oneirical stuff or, it is an absolutely consciousnessbal or awarenessbal stuff which is made of a condensed, compressed, compacted or congealed section, segment, part or portion of consciousness or awareness of cosmic space aka God, Brahman, Creator or Maker of cosmos. This section, segment, part or portion of consciousness or awareness of cosmic space aka God, Brahman, Creator or Maker of the cosmos has been condensed, compressed, compacted or congealed by the latter i.e., has been condensed, compressed, compacted or congealed by cosmic space aka God, Brahman, Creator or Maker of cosmos through the act of daydreaming or oneiricking on its part, nothing more nothing less.

This extreme-ground-level cosmology or, this extreme-end-point cosmology i.e., this cosmological truth is eternal in the manner cosmic space aka, God, Brahman, Creator or Maker of the cosmos is.

All the embodied human consciousnesses of planet earth who label, tag or dub the day-dreamal or oneirical stuff composed cosmos or, the consciousnessbal or awarenessbal stuff composed cosmos as being one hundred percent actual, factual, real, genuine or authentic or, as being one hundred percent material, substantial, physical, objective or concrete and as a consequence also claim that the cosmos is not made of day-dreamal or oneirical stuff or, consciousnessbal or awarenessbal stuff of cosmic space aka God, Brahman, Creator or Maker of the cosmos and instead is made of an actual, factual, real, genuine or authentic stuff called physical, objective or concrete matter which originated or emanated from an infinitesimally small, infinitely hot and infinitely dense physical thing called *singularity* or *cosmic egg* some 13.7 billion light years ago following a physical event called Big Bang which took place inside this physical but infinitesimally small, infinitely hot and infinitely dense *singularity or cosmic egg* under the influence of infinite physical temperature and infinite physical density present inside this physical *singularity or cosmic egg,* do not have their being outside this day-dreamal or oneirical stuff composed cosmos or, do not have their being outside this consciousnessbal or awarenessbal stuff composed cosmos.

Instead, their being is inside this day-dreamal or oneirical stuff composed cosmos or, inside this consciousnessbal or awarenessbal stuff composed cosmos.

Consequently, they are part and parcel of this day-dreamal or oneirical or, consciousnessbal or awarenessbal stuff composed cosmos which has been created or forged by cosmic space aka God or Brahman through the instrumentality of daydreaming or oneiricking on its part.

Since embodied human consciousnesses of planet earth are part and parcel of the day-dreamal or oneirical or, consciousnessbal or awarenessbal stuff composed cosmos which has been created or forged by cosmic space aka God or Brahman through the instrumentality of daydreaming or oneiricking on its part, the embodied human consciousnesses of planet earth do not have empirical experience of any other kind of cosmos against which they can judge whether the day-dreamal or oneirical or, the consciousnessbal or awarenessbal cosmos of which they are a constituent part, is really material, substantial, physical, objective or concrete in nature or, merely day-dreamal or oneirical in nature.

What has been said above needs to be expressed in another way in order to make it absolutely explicit for everyone. Even though the embodied human consciousnesses of planet earth, stereotype or brand the cosmos which they inhabit, as being material, substantial, physical, objective or concrete in nature but the supreme truth is that or, the eternal cosmological truth is that or, the extreme-ground-level or the extreme-end-point truth is that it is nothing of the sort.

Instead, the cosmos which the embodied human consciousnesses of planet earth inhabit, is a one hundred percent subjective construct, with no objective existence because it is a mere daydream or oneiric of cosmic space aka God, Brahman, Creator or Maker of the cosmos, as simple as that.

Since, in terms of absolute truth, the cosmos which the embodied human consciousnesses of planet earth inhabit, is a mere day-dreamal or oneirical entity or, is a mere day-dreamal or oneirical construct because it is made of day-dream stuff or oneiric stuff of cosmic space aka, God, Brahman, Creator or Maker of the cosmos, all the sciences, technologies, arts, music and sports of embodied human consciousnesses of planet earth are also mere day-dreamal or oneirical in nature or, are also mere day-dreamal or oneirical constructs because they also are made of mere day-dream stuff or oneiric stuff of cosmic space aka God, Brahman, Creator or Maker of the cosmos.

Similarly, bodies of all the embodied human consciousnesses of planet earth are not material, substantial, physical, objective or concrete bodies as they have been led to believe.

Instead, their bodies are day-dreamal or oneirical in nature, or, their bodies are day-dreamal or oneirical constructs because they are made of day-dream stuff or oneiric stuff of cosmic space aka God, Brahman, Creator or Maker of cosmos.

Therefore, all the activities undertaken by the day-dreamal or oneirical bodies of all the embodied human consciousnesses of planet earth are day-dreamal or oneirical activities only and nothing else.

What has been said above can be put in another way.

Bodies of all the embodied human consciousnesses of planet earth are not material, substantial, physical, objective or concrete bodies as believed by them.

Instead, bodies of all the embodied human consciousnesses of planet earth are mere day-dreamal or oneirical constructs because they all are made of mere day-dream stuff or oneiric stuff of cosmic space, aka, God, Brahman, Creator or Maker of the cosmos.

Therefore, all the activities undertaken by the day-dreamal or oneirical bodies of all the embodied human consciousnesses of planet earth are mere day-dreamal or oneirical activities and nothing less.

SUBJECTIVE AND OBJECTIVE REALITIES CONFRONTING EMBODIED HUMAN CONSCIOUSNESSES.

QUESTION.

There are words in the dictionary such as "empirical", "experiential", "existential", "observational" and "posteriori". What do they all mean?

ANSWER.

They are all synonymous terms. Thus, they all have the same meaning.

They denote deduction of theories from actual facts. That is to say, they refer to that, which is factual, actual, or real in the cosmos, or that, whose existence is verifiable by observation or experience in the cosmos rather than by a mere theory or, by a mere logic or by mere mathematic or arithmetic. In other words, these terms mean that, which can be observed, seen, and experienced directly or first-hand in the cosmos, or that, which is hands-on in the cosmos, or that, which is involved in active personal participation in the cosmos, or that, which is an experiential truth in the cosmos, or that, which is in practical, pragmatic, or in applied form in the cosmos, or that, with which one can do experiments, or, on which one can perform or set-up experiments, or, which can be subjected to experiments in the laboratories of the cosmos.

There are two fundamental kinds of "empirical", "experiential", "existential", "observational" or"posteriori" truths in the cosmos. These are :-

A. *"Empirical", "experiential", "existential", "observational" or "posteriori" truths relating to*

"OBJECTIVE REALITIES" *of the cosmos.*

B. *"Empirical", "experiential", "existential", "observational" or "posteriori" truths relating to* "SUBJECTIVE REALITIES" *of the cosmos.*

Let one deal with them one by one.

A. *What is an empirical, posteriori or experiential reality in the WORLD OF OBJECTIVE or MATERIAL SCIENCES which deal exclusively with OBJECTIVE or MATERIAL REALITIES of the cosmos?*

An empirical, posteriori or experiential reality *in the WORLD OF OBJECTIVE or MATERIAL SCIENCES* is a truth or fact which is confirmable, provable, testable or verifiable by physical sense perception, by physical sense observation, or, by physical sense experience rather than by mental or consciousnessbal logic alone or by mental or consciousnessbal theory alone or by mental or consciousnessbal mathematic or arithmetic alone.

An empirical, posteriori or experiential reality *in the WORLD OF OBJECTIVE or MATERIAL SCIENCES* is experienced by physical senses, that is to say, an empirical, posteriori or experiential reality *in the WORLD OF OBJECTIVE or MATERIAL SCIENCES* can be physically seen, observed or perceived by the physical senses. It is not purely a product of mental or consciousnessbal theory, logic, mathematic or arithmetic.

The existence of an empirical, posteriori or experiential reality *in the WORLD OF OBJECTIVE or MATERIAL SCIENCES* becomes confirmed only if it can be physically seen, observed, perceived or experienced via the media of the physical senses.

The veracity of the existence of an empirical, posteriori or experiential truth *in the WORLD OF OBJECTIVE or MATERIAL SCIENCES* is judged on the basis that it can be physically seen, observed or experienced via the media of the physical senses.

An empirical, posteriori or experiential truth *in the WORLD OF OBJECTIVE or MATERIAL SCIENCES* is capable of being verified by physical observations or experiments.

An empirical, posteriori or experiential truth *in the WORLD OF OBJECTIVE or MATERIAL SCIENCES* is verifiable by physical experiments and physical observation relating to them.

The existence of an empirical, posteriori or experiential truth *in the WORLD OF OBJECTIVE or MATERIAL SCIENCES* is based on practical experience of physical kind.

That is to say, the existence of an empirical, posteriori or experiential truth *in the WORLD OF OBJECTIVE or MATERIAL SCIENCES* is based on actual doing something of physical kind by use of some kind of physical gadget or physical instrument rather than based on mental or consciousnessbal theory, ideas, mathematic, or, arithmetic.

An empirical, posteriori or experiential knowledge *in the WORLD OF OBJECTIVE or MATERIAL SCIENCES* is derived from physical sense experience rather than by mental or consciousnessbal logic, theory, mathematic, or, arithmetic.

An empirical, posteriori or experiential evidence or knowledge *in the WORLD OF OBJECTIVE or MATERIAL SCIENCES* is defined as an evidence or knowledge which is obtained by physical sense perception or physical sense experience. It is an evidence or knowledge acquired by means of physical senses, particularly by physical observation and physical experimentation.

The definition of the words "empirical", "posteriori" or "experiential" *in the WORLD OF OBJECTIVE or MATERIAL SCIENCES* is something, that is based solely on physical experiment or physical experience.

The definition of the term "empirical knowledge", "posteriori knowledge" or "experiential knowledge" *in the WORLD OF OBJECTIVE or MATERIAL SCIENCES* is the following:-
"It is that kind of knowledge which is derived from physical investigation, observation, experimentation, or physical experience, as opposed to theoretical knowledge based on logical, mathematical, or, arithmetical assumptions of mental or consciousnessbal kind".

The word "empirical", "posteriori" or "experiential" *in the WORLD OF OBJECTIVE or MATERIAL SCIENCES* means "physical experience". Therefore, a knowledge gained by physical experience is called an "empirical knowledge", "posteriori knowledge" or "experiential knowledge".

An "empirical knowledge" or "experiential knowledge" *in the WORLD OF OBJECTIVE or MATERIAL SCIENCES* is also called a "posteriori" knowledge" but "posteriori knowledge" of the physical kind and not of consciousnessbal kind.

Let one make clear that consciousnessbal kind of "posteriori knowledge" aka "empirical knowledge" or "experiential knowledge" belongs to the WORLD OF

SUBJECTIVE REALITIES and thus, to those *DISCIPLINES, FIELDS, AREAS OF STUDY, or, BRANCHES OF KNOWLEDGE which deal exclusively with* SUBJECTIVE REALITIES *of the cosmos.*

The word "posteriori" simply means "experiences" without implying whether "experiences" or "posteriori" in question are of physical kind or, one hundred percent, of consciousnessbal kind.

However, one word of extreme caution here to all, which is vital. This caution is as follows :-

Ultimately, or, in the final analysis, all human "experiences" or "posteriori", irrespective of whether they have been collected or stockpiled by embodied human consciousnesses indirectly via the involvement or intervention of objective realities of the cosmos or, directly via the media of subjective realities of the cosmos such as human pains and joys of one hundred percent, mental or consciousnessbal variety i.e., human pains and joys taking birth in embodied human consciousness without the involvement or intervention of any kind of physical objects, and all other human emotions taking birth in embodied human consciousness without the involvement or intervention of any kind of physical objects, plus all human thoughts, ideas, desires and ambitions taking birth in embodied human consciousnesses without the involvement or intervention of any kind of physical objects, human day-dreams i.e., human oneirics or human-dreaming while awake, taking birth in embodied human consciousnesses without the involvement or intervention of any kinds of physical objects, human night-dreams i.e., human night-irics or human-dreaming while human consciousnesses are in their dream-sleep-states, (these too take birth in embodied human consciousnesses without the involvement or intervention of any kinds of physical objects) and last, but not least, human consciousnesses on one hand and ubiquitous and infinite field of consciousness aka cosmic space on the other; are all such "experiences" or "posteriori" which are all gathered, garnered, amassed or accumulated by embodied human consciousnesses only and not by their bodies because without the presence of consciousness in the body, the latter and all its five physical sense organs are one hundred percent useless.

The knowledge gained by experiences, that is to say, knowledge gained from "observed facts" - ("observed facts" are "experiences") - is known as empirical knowledge or posteriori knowledge.

"Experiences" or "observed facts" are of two kinds in the cosmos, namely :-

1. "Experiences" or "Observed Facts" of *"physical kind"* on one hand.

And

2. "Experiences" or "Observed Facts" of absolutely or one hundred percent *"consciousnesses kind"* on the other.

Let one recapitulate or go over once again what one means when one says :- "Experiences" or "Observed Facts" of absolutely or one hundred percent *"consciousnessbal variety"*.

"Experiences" or "observed facts" of absolutely or one hundred percent of consciousnessbal variety relate to those "experiences" or "observed facts" which exclusively belong to the world of subjective realities of the cosmos such as human pains and joys of one hundred percent, mental or consciousnessbal variety i.e., human pains and joys taking birth in embodied human consciousness without the involvement or intervention of any kind of physical objects, and all other human emotions taking birth in embodied human consciousness without the involvement or intervention of any kind of physical objects, plus all human thoughts, ideas, desires and ambitions taking birth in embodied human consciousnesses without the involvement or intervention of any kind of physical objects, human day-dreams i.e., human oneirics or human-dreaming while awake, taking birth in embodied human consciousnesses without the involvement or intervention of any kinds of physical objects, human night-dreams i.e., human night-irics or human-dreaming while human consciousnesses are in their dream-sleep-states, (these too take birth in embodied human consciousnesses without the involvement or intervention of any kinds of physical objects) and last, but not least, human consciousnesses on one hand and the ubiquitous and infinite field of consciousness aka cosmic space on the other.

QUESTION.

What is a Priori knowledge, or a Deductive knowledge or a Inferential knowledge?

ANSWER.

Before one answers the above question :- "What is a Priori knowledge or a Deductive knowledge or a Inferential knowledge"? let one explain, once more, what an empirical knowledge or posteriori knowledge is?

An empirical knowledge or posteriori knowledge is garnered or gathered by the physical sense perception, observation, or, by physical sense experience.

A Priori knowledge, Deductive knowledge or an Inferential knowledge, on the other hand, is defined as that knowledge which is garnered or gathered by reason or logic alone or by theory, mathematic or arithmetic alone.

Priori knowledge on one hand and posteriori knowledge on the other are frequently contrasted with each other in order to distinguish, one from the other.

To reiterate.

A knowledge, garnered or gathered by reason alone or logic alone or by theory alone or mathematic alone or arithmetic alone, is called a priori knowledge or a deductive knowledge or an inferential knowledge.

B. *What is an empirical, posteriori or experiential reality in those DISCIPLINES, FIELDS, AREAS OF STUDY, or, BRANCHES OF KNOWLEDGE which deal exclusively with* SUBJECTIVE REALITIES *confronting embodied human consciousnesses of planet earth?*

Human pains and joys, human thoughts, ideas, desires and ambitions, human day-dreams i.e., human oneirics or human-dreaming while awake, human night-dreams i.e., human night-irics or human-dreaming while human consciousnesses are in their dream-sleep-states, and last, but not least, human consciousnesses on one hand and the ubiquitous and infinite field of consciousness aka cosmic space on the other, are all subjective realities of the cosmos but, at the same time, they all are also empirical, posteriori or experiential realities of the cosmos, that is to say, they all are also such realities of the cosmos which are experienced by all embodied human consciousnesses directly, personally or unmediated or, experienced by all embodied human consciousnesses on first-hand basis or hands on basis or, experienced by all embodied human consciousnesses from the original source. Therefore, all the above listed, counted or enumerated subjective realities of the cosmos are designated as empirical, posteriori or experiential realities which are on a par with or in a class with the one and only objective reality of the cosmos, namely, physical matter.

To paraphrase.

Human pains of both physical and mental or, physical and consciousnessbal varieties, human joys and all other human emotions plus all human thoughts, ideas, desires and ambitions, human day-dreams i.e., human oneirics or human-dreaming while awake, human night-dreams i.e., human night-irics or human-dreaming while human consciousnesses are in their dream-sleep-states, and last, but not least, human consciousnesses on one hand and the ubiquitous and infinite field of consciousness aka cosmic space on the other, are all subjective realities but at the same time they are all also empirical, posteriori or experiential realities that is to say, they all are also such realities which are experienced by all embodied human consciousnesses directly, personally or unmediated or, experienced by all embodied human consciousnesses on first-hand basis or hands on basis or, experienced by all embodied human consciousnesses from the original source. Therefore, all the above listed, counted or enumerated subjective realities are designated empirical, posteriori or experiential realities which are on a par with or in a class with, the one and only objective reality of the cosmos, namely, physical or concrete matter.

With regards to the ubiquitous and infinite field of consciousness aka cosmic space, one will like to add that this ubiquitous and infinite field of consciousness aka cosmic space provides a 3-D or three-dimensional, consciousnessbal space, room, or territory for the spatial or territorial placement and existence of the 3-D or three-dimensional, objective but quintessentially, consciousnessbal in nature, physical matter so that the latter can take birth inside the ubiquitous and infinite field of consciousness aka cosmic space through the instrumentality of daydreaming or oneiricking on its part and become a vehicle to create variety, diversity, multiplicity, or heterogeneity in the nondescript or featureless milieu, backdrop or setting of the ubiquitous and infinite field of consciousness aka cosmic space.

All that has been said above can be put in another way.

All the items listed below are subjective realities but at the same time they all are also empirical, posteriori or experiential realities that is to say, they all are realities experienced by all embodied human consciousnesses directly, personally or unmediated or, experienced by all embodied human consciousnesses on a first-hand basis or hands on basis or, experienced by all embodied human consciousnesses from the original source. Therefore, all these subjective realities are designated empirical, posteriori or experiential realities which are on a par with or in a class with the one and only objective reality of the cosmos, namely, physical or concrete matter.

The list of all the subjective but also empirical, posteriori or experiential realities of the cosmos is as follows.

All human pains, both physical and mental or, physical and consciousnessbal varieties, all human joys and all other human emotions plus all human thoughts, ideas, desires and ambitions, all human day-dreams i.e., all human oneirics or human-dreaming while awake, all human night-dreams i.e., all human night-irics or human-dreaming while human consciousnesses are in their dream-sleep-states, and last, but not least, all human consciousnesses on one hand, and the one and only, plus the ubiquitous and infinite field of consciousness aka cosmic space on the other.

Incidentally, the one and only plus the ubiquitous and infinite field of consciousness aka cosmic space aka God or Brahman or the Creator of cosmos affords a 3-D or three-dimensional, consciousnessbal space, room or territory to the 3-D or three-dimensional, objective, but quintessentially consciousnessbal in nature, physical matter so that the latter, namely, the 3-D or three-dimensional, objective, but quintessentially consciousnessbal in nature, physical matter can take birth inside the ubiquitous and infinite field of consciousness aka cosmic space aka God or Brahman, through the instrumentality of daydreaming or oneiricking on its part and become a vehicle to create variety, diversity, multiplicity, or heterogeneity in the nondescript or featureless milieu, backdrop or setting of the ubiquitous and infinite field of consciousness aka cosmic space aka God or Brahman.

As already said above, there is only thing in the cosmos which is an objective reality in mankind's fully awake state & that is physical matter; it quintessentially is consciousnessbal in nature and anything which is quintessentially consciousnessbal in nature is subjective fundamentally even though it gives the impression, sense, or feeling to consciousness in question as if it were objective, physical, material, substantial or concrete in texture.

With the sole exception of physical matter, the rest of the items listed above, which are either *sensed*, on one hand, i.e., human consciousness and its emotions, thoughts, ideas, desires and the like, or, *seen* i.e., cosmic space, on the other, all of whom are directly or on first-hand or hands-on basis, experienced by all embodied human consciousnesses of the planet earth in their fully-awake or wide-awake state, are subjective realities.

However, as explained above, even all the subjective realities of the cosmos, which are either merely sensed, for example, human consciousness and its emotions, thoughts, ideas, desires and the like, on one hand, or, are merely seen i.e., cosmic space, on the other, by all embodied human consciousnesses of planet earth in their fully-awake or wide-awake state, are all empirical, posteriori or experiential realities or such realities which are experienced by them in their fully-awake or wide-awake state, directly, personally or unmediated or, experienced by them on first-hand basis or hands on basis or, experienced by them from the original source. Therefore, all subjective realities of the cosmos which are either sensed or seen by all embodied human consciousnesses of planet earth in their fully-awake or wide-awake state, are rightfully described as empirical, posteriori or experiential realities which are on a par with or in a class with the one and only objective reality of the cosmos, namely, physical or concrete matter.

Therefore, embodied human consciousnesses should not be too "gung-ho" or too keyed up, fired up, dedicated, devoted, jealous, aggressive, ardent, eager, enthusiastic, energetic, enthused or excited about the objective matter i.e., physical matter of the cosmos and the myriad items made or constructed out of the objective matter of the cosmos as is their wont, habit or conditioning at present because there is more to embodied human consciousnesses' existence or presence in the cosmos than objective or physical matter alone.

In any case, the foundation, bed-rock or substratum, on which the edifice or the monument of the one and only objective reality of the cosmos, - all other's realities of the cosmos are subjective realities - is standing, is the unique plus ubiquitous and infinite subjective reality of the cosmos which, on account of the cosmic delusion or cosmic ignorance of embodied human consciousnesses of planet earth, is labeled or tagged by them as cosmic space but which in fact is the one and only and incredible plus ubiquitous and infinite field of consciousness of God, Brahman, Creator, Maker or Progenitor of the cosmos, the cosmos of which both matter and consciousness are the basic, constituent parts, out of which has come the mind-boggling variety or diversity, witnessed today in the cosmos.

This one and only and incredible plus ubiquitous and infinite field of consciousness of God, Brahman, Creator, Maker or Progenitor of the cosmos - which on account of the cosmic delusion or cosmic ignorance of embodied human consciousnesses of planet earth is wrongly identified or tagged by them as cosmic space in their fully-awake or wide-awake state - has given birth or nativity to cosmos within itself through the process of daydreaming or oneiricking or, through the process of mental imaging or consciousnessbal imaging on its part and nothing else.

All objective or material sciences make "too much" or "over the top", "hoo-ha", or "song and dance" about the empiricism or the empirical nature of physical matter of the cosmos and all the empirical sciences which have evolved or emanated from this physical matter through no doubt very laborious and painstaking or, through no doubt very challenging and arduous, scientific efforts of countless, brilliant embodied human consciousnesses of planet earth.

Unfortunately many members of the scientific community, who no doubt are doing praiseworthy empirical work in their chosen fields of objective sciences, are in the regrettable habit of disseminating misleading information that the empiricism is the hallmark of objective sciences only and that there exists no empiricism in the world of subjective realities of the cosmos nor is there any empiricism in the disciplines, fields, areas of study or branches of knowledge which are connected with the subjective realities of the cosmos.

To paraphrase.

Many distinguished scientists who are devoting their lives to the world of physical, material or objective sciences, give the erroneous impression that their scientific world, which deals exclusively with physical objects, is the only world which is based on empiricism or empirical evidence.

However, the truth is very different because the vast and very varied world of subjective realities of the cosmos and the phenomenal amount of work which has been undertaken by countless brilliant embodied human consciousnesses of planet earth over the past many millenniums with regards to these subjective realities of the cosmos, undoubtedly also have as much footing in the sanctum santorum of empiricism as the objective sciences of the planet earth have.

In their very laudable effort to highlight the sacrosanctness of empiricism in the various fields of objective sciences, material scientists, barring few exceptions, have regretfully overlooked or rather downplayed the empiricism or the empirical nature of subjective realities of the cosmos and the very vast amount of work which has been undertaken by numerous, extremely gifted embodied human consciousnesses of planet earth over the past many millenniums with respect to them. Their countless works and discoveries, made in the fields of subjective realities of the cosmos, are as empirical or experiential in nature as the vast amount of work connected with the objective or physical matter of the cosmos are.

The Cosmos is one single or unitary whole but it consists of two halves namely objective and subjective or, physical and consciousnessbal. They are both inseparably joined together to make it one single or unitary whole.

Therefore to label one i.e., physical matter aka objective realities, as empirical, experiential, existential, experimentable, testable, verifiable, confirmable, observational, or data-based and the other i.e., human consciousnesses, cosmic space, human emotions, human ideas, human thoughts and the like aka subjective realities, as non-empirical, non-experiential, non-existential, non-observational, non-data-based, unexperimentable, un-testable, un-verifiable, un-confirmable, or, suppositional, conjectural, hypothetical, theoretical, abstractive, speculative, metaphysical, or divinatory and that too, without applying their consciousness or mind wholeheartedly or earnestly to it or without giving wholehearted or earnest attention to it, plus without exploring the world of subjective realities unbiasedly or dispassionately, is irrational, to say the least.

This irrationality of those embodied human consciousnesses who are engaged with objective matter alone plus its sciences, is handicapping them terribly from discovering the "unitary truth" of the cosmos.

This "unitary truth" of the cosmos is supporting the whole cosmos on one hand and is also its foundation, bed-rock or substratum, on the other.

The name of this "unitary truth" of the cosmos is cosmic space.

This cosmic space aka the "unitary truth" of the cosmos, is seen by all embodied human consciousnesses in their fully-awake or wide-awake state via the media of their physical eyes, every moment of their existence in the cosmos.

All embodied human consciousnesses must remember the very famous aphorism, truism, maxim or dictum that the "unitary truth" of the cosmos aka the "absolute truth" or the "supreme truth" of the cosmos which stands behind the cosmos in a partially veiled form, when unveiled or discovered by them, after becoming free of their cosmic delusion or cosmic ignorance, will be supremely simple, in fact as simple as the back of one's hand and will therefore be accessible even to a child without involving too much effort or exertion on anyone else's part.

CONSCIOUSNESSBAL SENSES USED IN DREAM SLEEP STATE VS PHYSICAL SENSES USED IN WAKEFUL STATE

Embodied human consciousnesses of planet earth possess five sensory portals of the physical kind and the same number of the consciousnessbal kind.

Let one explain what one means.

The five sensory portals of the physical kind, namely, those of eyes, ears, nose, tongue, and skin of the embodied human consciousnesses operate exclusively during their wakeful state. They do not function during their dream sleep state.

These five sensory portals, function via a complex physical medium, consisting of their connecting physical nerves and then the physical spinal cord and finally the physical brain.

It is in the brain that all the sensory inputs, received by the five sensory portals of the body, from the outside world are interpreted by the embodied human consciousness in order to make sense of the outside world, which the embodied human consciousness in question observes, perceives, and experiences in its wakeful state through the media of these five sensory portals of the physical body.

Let one remind oneself that the outside world, with respect to the wakeful state of an embodied human consciousness, is called the material, substantial, physical, objective or concrete world.

Incidentally, the role of the physical brain in the human body is that of an *antenna* to start with, an *antenna which* gathers, collects, draws-in or sucks-in consciousness from the ubiquitous and infinite ocean of consciousness aka cosmic space.

Following its initial role of an *antenna* to start with, - an *antenna which* gathers, collects, draws-in or sucks-in consciousness from the ubiquitous and infinite ocean of consciousness aka cosmic space, - the physical brain of the human body then also takes on the role of the storage, depot, or vault of consciousness plus the role of the distribution centre of consciousness from whence the consciousness is then distributed or dispersed throughout the body.

As said earlier, the five sensory portals of the physical kind, namely, those of the eyes, ears, nose, tongue, and skin of embodied human consciousnesses, operate exclusively during their wakeful state but do not operate during their dream sleep state because their dream sleep state is an extremely uncanny, unusual, mystifying or weird state of sleep on many counts.

The extremely uncanny, unusual, mystifying or weird state of sleep called dream sleep state has been, deliberately or on purpose, afforded to human consciousness by cosmic space aka the creator of the *physical matter* aka the creator of physical cosmos, with a very specific goal in mind. The nature of this specific goal will become clear as the whole narrative of this chapter proceeds towards completion.

The many strange, surreal, eerie, or "other-worldly" aspects of dream sleep state of the human consciousness are as described below.

Firstly, during its dream sleep state, a human consciousness, - unlike during its deep sleep state, - even though in sleep, remains a conscious being, in fact as conscious as it was during its wakeful state.

That is to say, during its dream sleep state, a human consciousness, even though asleep, is not unconscious in the manner it is in its deep sleep state. This is a highly unusual event, to say the least or to put it mildly, unusual because, to be asleep and yet remaining conscious, seems like an oxymoronic expression, and yet this is a weird fact which happens in the life of the human consciousness, a weird fact which takes place daily when a human consciousness goes to fulfill its business of sleep each night.

Secondly, during the dream sleep state, a human consciousness, even though conscious, (in fact as conscious as it was during its wakeful state), is magically unaware of the existence of the physical, material, substantial, concrete or objective world of which it was aware in its wakeful state and which was its *"HEART and SOUL"* or, should one say, which was its *only "DEITY OF WORSHIP"* during that state.

Along with the incredible experience of unawareness of the physical world during its dream sleep state, - (dream sleep state during which, a supposed to be asleep human consciousness, remains aware or conscious, in fact as aware or conscious as it was during its wakeful state) -, two additional but very puzzling experiences, are handed out to human consciousness by cosmic space aka the Creator of *physical matter* aka the Creator of physical cosmos.

To start with, a human consciousness, even though aware or conscious in its dream sleep state, - (in fact as aware or conscious as it was during its wakeful state), - becomes unaware or unconscious of its physical body, its physical body of which it was aware or conscious every moment when it was in its wakeful state.

Thus, during the short period of its dream sleep state, human consciousness, temporarily becomes a mind-boggling or mind-blowing bodiless consciousness in the manner of its spring, source or, fountainhead, namely, cosmic space aka the creator of the *physical matter* aka the creator of the physical cosmos is.

Secondly, human consciousness, during its awesome dream sleep state, - when it temporarily exists as mind-boggling or mind-blowing bodiless consciousness, - (in stark contrast, during its wakeful stake, human consciousness exists as embodied consciousness), - encounters, observes, watches, sees, perceives or experiences a magical world.

This magical world, encountered, watched, observed, seen, perceived and experienced by human consciousness as a "bodiless being", during its awesome dream sleep state, is labeled or tagged the dream sleep world, cosmos or universe, even though at the time of watching that magical world, cosmos or universe during its dream sleep state, this very same human consciousness was unaware that the latter world, cosmos or universe was a mere dream world, cosmos or universe and not a physical world, cosmos or universe.

Identical, similar, alike or exactly the same is the state of affairs, situation, condition, case, or ball game with regards to the world, cosmos or universe which human consciousness encounters, observes, watches, sees, perceives, experiences and takes part in during its wakeful state.

The world, cosmos or universe which human consciousness encounters, observes, watches, sees, perceives, experiences and takes part in during its wakeful state, is also a dream-world but not a dream-sleep-world in the manner the dream-sleep-world is; the dream-sleep-world which human consciousness encounters, watches, observes, sees, perceives, experiences and takes part in, during its dream sleep state.

Instead, the world, cosmos or universe which human consciousness encounters, observes, watches, sees, perceives and experiences during its wakeful state is a daydream world, cosmos or universe or an oneirickal world, cosmos or universe.

This daydream world, cosmos or universe or oneirickal world, cosmos or universe, which human consciousness encounters, observes, watches, sees, perceives and experiences during its wakeful state has been created by cosmic space aka the creator of *physical matter* aka the creator of physical cosmos through the very ordinary or common-'o'-garden activity called daydreaming or oneiricking on its part and nothing else.

The dream sleep world, cosmos, or universe, encountered, watched, observed, seen, perceived and experienced by the awesome "bodiless human consciousness" during its awesome dream sleep state is such a magical world, cosmos or universe which, whilst being encountered, watched, observed, seen, perceived and experienced by the awesome "bodiless counterpart or twin" of human consciousness during its awesome dream sleep state on one hand, and being not only encountered, watched, observed, seen, perceived and experienced but also "taken part in it", by the awesome "embodied counterpart or twin" of the same human consciousness on the other, was accepted by the awesome "*both counterparts* or *both twins"* of one and the same human consciousness as being real or genuine or, was being accepted by "them both" as being "in the same class as" or, "on a par with" the magical world, cosmos or universe which the same human consciousness encounters, observes, watches, sees, perceives, experiences and takes part in, when it is in its incredible wakeful state as well as in its incredible embodied state, both at one and the same time.

Hence, as said earlier and here one quotes :- "Identical, similar, alike or exactly the same is the state of affairs, situation, condition, case, or ball game with regards to the world, cosmos or universe which human consciousness encounters, observes, watches, sees, perceives, experiences and takes part in during its wakeful state.

This world, cosmos or universe which human consciousness encounters, observes, watches, sees, perceives, experiences and takes part in during its wakeful state, is also a dream world, cosmos or universe but not a dream sleep world, cosmos or universe in the manner the dream sleep world, cosmos or universe is, the dream sleep world, cosmos or universe which human consciousness encounters, watches, observes, sees, perceives, experiences and takes part in, in its dream sleep state. Instead, it is a daydream world, cosmos or universe or oneirickal world, cosmos or universe, created by cosmic space aka the creator of *physical matter* aka the creator of physical cosmos, through the very ordinary or common-'o'-garden activity of daydreaming or oneiricking on its part and nothing else.

To repeat.

Hence, as said earlier and here one quotes :- "Identical, similar, alike or exactly the same is the state of affairs, situation, condition, case, or ball game with regards to the world, cosmos or universe which human consciousness encounters, observes, watches, sees, perceives, experiences and takes part in during its wakeful state. This world, cosmos or universe which human consciousness encounters, observes, watches, sees, perceives, experiences and takes part in during its wakeful state, is also a dream world, cosmos or universe but a daydream world, cosmos or universe or oneirickal world, cosmos or universe, created by cosmic space aka the creator of *physical matter* aka the creator of physical cosmos through the very ordinary or common-'o'-garden activity of daydreaming or oneiricking on its part and nothing else".

One very important point to spot or take notice of here in connection with the subject matter one is dealing with at present is the following :-

The one and the same human consciousness exists for some hours each day as bodiless consciousness and for some hours each day as embodied consciousness.

To start with, in its awesome dream sleep state human consciousness exists in its two Avatarik forms or in its two Incarnational forms.

In its awesome dream sleep state, human consciousness exists on one hand, as an awesome bodiless consciousness and on the other, as an awesome embodied consciousness, at one and the same time.

The *bodiless* avatarik or incarnational form of human consciousness which is extant or present in the latter's dream sleep state, simply observes, watches, sees, perceives or experiences the magical world called the dream sleep world but it does not take part in it.

In contrast, the *embodied* avatarik or incarnational form of human consciousness which is extant or present in the latter's dream sleep state, not only observes, watches, sees, perceives or experiences the magical world called the dream sleep world but also takes part in it.

The *embodied* avatarik or incarnational form of human consciousness which is extant or present in the latter's dream sleep state and who not only encounters, observes, watches, sees, perceives or experiences the magical world called the dream sleep world but also takes part in it, dies when the magical world called the dream sleep world dies.

In contrast, the *bodiless* avatarik or incarnational form of human consciousness which was extant or present in the latter's dream sleep state and who simply watched, observed, saw, perceived and experienced the magical world called the dream sleep world but did not take part in it, wakes up after the end of sleep of human consciousness and once again becomes aware that it has a body or once again becomes aware that it is embodied and is no longer bodiless in the manner it was during its dream sleep state.

This *embodied* avatarik or incarnational form of human consciousness which is extant or present in the latter's wakeful state, then restarts not only watching, observing, seeing, perceiving and experiencing its awesome wakeful state world, cosmos or universe but also restarts taking part in it, because it has no alternative or no option or choice in this matter.

The above narrated, repetitive or recurring routine, pattern, or drill which human consciousness has to go through daily, or each twenty four hours, consisting of being embodied in its wakeful state on one hand and of being both bodiless and embodied at one and the same time in its dream sleep state on the other, followed by the death of its embodied twin (which existed in its dream sleep state) on waking up from sleep plus conversion of the bodiless status of its other twin (which existed in dream sleep state) into embodied-ness once again after waking up from sleep, is melodiously and picturesquely described in the Adwait-Vedantic parlance as the "Cosmic Dance of Shiva" or the "Cosmic Dance of Rudra" or the "Tandava Dance of Shiva" or the "Tandava Dance of Rudra".

Paraphrased in English, the expression "Cosmic Dance of Shiva (Rudra)" or the "Tandava Dance of Shiva (Rudra)", describes the above narrated, daily, twenty-four hourly, or everyday repetitive or recurring routine, pattern or drill of human consciousness as the "Cosmic or the Tandava Dance of C*reation & Destruction of* P*hysical Matter",* which has been choreographed, directed, and enacted by none other than the "The Supreme C*reator and Destroyer of physical matter"* aka cosmic space. And this cosmic space is nothing but the ubiquitous and infinite field of consciousness of this "Supreme C*reator and Destroyer of physical matter".*

Thus, embodisization or embodied-ness of human consciousness is not permanent.

Embodisization or embodied-ness of a human consciousness alternates, pendulates, see-saws, yo-yos, takes turn or changes back and forth with its non-embodisization or bodiless-ness each day or each twentyfour hours, just as the state of its wakefulness alternates, pendulates, see-saws, yo-yos, takes turn or changes back and forth with the state of its dream-sleepness and deep-sleepness each day or each twentyfour hours.

What has been said above needs to be repeated so that it becomes etched in one's consciousness.

The one and the same human consciousness exists some hours each day as bodiless consciousness and some hours each day as embodied consciousness.

To start with, in its awesome dream sleep state, the one and the same human consciousness exists in two avatarik or incarnational forms.

One of these avatarik or incarnational form exists in an incredible bodiless form whereas the other exists in an incredible embodied form.

The bodiless avatarik or incarnational form of human consciousness, which is extant or present in the latter's dream sleep state, simply observes, watches, sees, perceives or experiences the magical world called the dream sleep world but it does not take part in it.

In contrast, the embodied avatarik or incarnational form of human consciousness, which is extant or present in the latter's dream sleep state, not only observes, watches, sees, perceives or experiences the magical world called the dream sleep world but also takes part in it.

The embodied avatarik or incarnational form of human consciousness, extant or present in the latter's dream sleep state, which not only encounters, observes, watches, sees, perceives or experiences the magical world called the dream sleep world but also takes part in it, dies when the magical world called the dream sleep world dies.

In contrast, the bodiless avatarik or incarnational form of human consciousness, extant or present in the latter's dream sleep state, which simply observes, watches, sees, perceives or experiences the magical world called the dream sleep world but does not take part in it, wakes up after the end of sleep of human consciousness and once again becomes aware that it has a body or once again becomes aware that it is embodied and is no longer bodiless in the manner it was during its dream sleep state.

The embodied, avatarik or incarnational form of human consciousness which is extant or present during the latter's dream sleep state then restarts with not only watching, observing, seeing, perceiving and experiencing its awesome wakeful state world, cosmos or universe but also restarts taking part in it because it has no alternative or, it has no option or choice in this matter.

Thus, embodisization or embodied-ness of a human consciousness is not permanent.

The embodisization or embodied-ness of a human consciousness alternates, pendulates, see-saws, yo-yos, takes turn or changes back and forth with its non-embodisization or bodiless-ness each day or each twenty-four hours, just as the state of its wakefulness alternates, pendulates, see-saws, yo-yos, takes turn or changes back and forth with the state of its dream-sleepness and deep-sleepness each day or each twenty-four hours.

The incredible creator of *physical matter* in the cosmos namely the ubiquitous and infinite field of consciousness aka cosmic space has deliberately or purposely sketched or chalked out the existence of human consciousness in the above described manner with the hope that one day human consciousness will rise above its cosmic delusion and utilise all its mental or consciousnessbal energy for the purpose of fully grasping the modus operandi by which the creation of *physical matter* has taken place in the cosmos.

While on the subject of the presence of *physical matter* in the cosmos, one must remind oneself that while the *consciousness* is the most important ingredient of the cosmos, *physical matter* is the second most important ingredient of the cosmos. These two have joined hands to constitute the cosmos.

Physical matter is a "created item" in the cosmos whereas consciousness is not. Consciousness is eternal. Being eternal, is the inherent nature of all consciousnesses. That is to say, being eternal, is the inherent nature of all consciousnesses present in the cosmos irrespective of whether the consciousness in question is the ubiquitous and infinite field of consciousness aka cosmic space aka god or, human consciousnesses.

Since *physical matter* is the second most important ingredient of cosmos (after *consciousness*), it needed to be created in the cosmos if birth of variety, diversity, multiplicity or heterogeneity had to take place inside the nondescript or featurelesssetting of the ubiquitous and infinite field of consciousness aka cosmic space. *Physical matter* has been created in the cosmos by the ubiquitous and infinite field of consciousness aka cosmic space aka god. The latter has created *physical matter* in the cosmos through the very ordinary or common-'o'-garden activity on its part called daydreaming or oneiricking.

The ubiquitous and infinite field of consciousness aka cosmic space aka god aka creator of *physical matter is "SWAYAMBHU".*

"Swayambhu" is a Sanskrit word which means " that who is Self-existent", or "that who is existent by its own accord, volition, free will, choice, discretion, desire or wish" or "that who has reality existence by its own accord or free will" or "that who is existing without having been created" or "that who is existing independent of any extrinsic sustaining force" or "that who is existing independently of other beings or causes".

Let us now come back to the topic or subject matter of dream sleep state of human consciousness which one was dealing with earlier.

The magnificent magical world, labeled as dream sleep world, which the *bodiless* avatarik form i.e., version of human consciousness, present in latter's dream sleep state, perceives and experiences but does not take part in it plus the *embodied* avatarik form i.e., version of human consciousness, present in the latter's dream sleep state who not only perceives and experiences this magical world i.e., dream sleep world but also takes part in it, is as actual-looking, factual-looking, real-looking, genuine-looking, authentic-looking plus material-looking, substantial-looking, physical-looking, objective-looking, concrete-looking and diverse-looking, varied-looking, variegated-looking or manifold-looking as well as, awe-inspiring or mind-blowing as the material, substantial, physical, objective or concrete world, cosmos or universe is, the material, substantial, physical, objective or concrete world, cosmos or universe which human consciousness as an embodied being encounters, watches, observes, sees, perceives or experiences during its awesome wakeful state.

Unfortunately, on account of its cosmic delusion, human consciousness, when it returns to wakefulness and therefore regains the awareness of its embodied-ness, disdainfully dismisses the actual-looking, factual-looking, real-looking, genuine-looking or authentic-looking plus material-looking, substantial-looking, physical-looking, objective-looking or concrete-looking as well as manifold or diverse looking magical world called the dream sleep world, (which it perceived and experienced as well as took part in it through its two incarnations or avatars during its dream sleep state the previous night) , as a mere "dream of its mind", "fantasy of its mind", "make-belief of its mind", "bubble of its mind", "vapour of its mind" or "trick of its mind".

By adopting the above described dismissive, disdainful or disparaging attitude vis-a-vis the magical world called the dream sleep world, the subconscious strategy, tactic or game plan in the mind of embodied human consciousness of the wakeful state is to convince itself that the actual-looking, factual-looking, real-looking, genuine-looking or authentic-looking plus material-looking, substantial-looking, physical-looking, objective-looking or concrete-looking as well as manifold or diverse looking magical world called the wakeful state world , which it perceives and experiences plus in which it takes part whenever it is in its wakeful state and of which its actual-looking, factual-looking, real-looking, genuine-looking or authentic-looking plus material-looking, substantial-looking, physical-looking, objective-looking or concrete-looking body is a part and parcel, is not like the magical world called the dream sleep world or, absolutely to the point, is not like the dream-stuff-composed-world called the dream sleep world which it perceived and experienced plus took part in, through its two avatars or incarnations during its dream sleep state the previous night or whenever.

The above described more favourable but undeserved treatment offered to the actual-looking, factual-looking, real-looking, genuine-looking or authentic-looking plus material-looking, substantial-looking, physical-looking, objective-looking or concrete-looking as well as manifold or diverse looking magical world called the wakeful state world, by the embodied human consciousness of the wakeful state as compared to the dream sleep world of its dream sleep state is illogical, to say the least.

Embodied human consciousness of the wakeful state, has no logical, convincing or, cogent, compelling or clear reason to support its more favourable but undeserved treatment of the actual-looking, factual-looking, real-looking, genuine-looking or authentic-looking plus material-looking, substantial-looking, physical-looking, objective-looking or concrete-looking as well as manifold or diverse looking magical world called the wakeful state world, which it perceives , experiences and which takes part in during its wakeful state, as compared to the dream-stuff-composed, *dream sleep world* which it perceives and experiences during its dream sleep state.

The dream-stuff-composed, *dream-sleep-world,* which human consciousness perceives and experiences and which it takes part in through its two avatars or incarnations during its dream sleep state, is as actual-looking, factual-looking, real-looking, genuine-looking or authentic-looking plus material-looking, substantial-looking, physical-looking, objective-looking or concrete-looking as well as manifold or diverse looking as *the actual-looking, factual-looking, real-looking, genuine-looking or authentic-looking plus material-looking, substantial-looking, physical-looking, objective-looking or concrete-looking as well as manifold or diverse looking world* which human consciousness perceives, experiences and takes part in during its wakeful state, which it calls the wakeful state world.

Both magical worlds, perceived, experienced and taken part in by human consciousness per twenty four hours, - namely, the one, perceived, experienced and taken part in during its dream sleep state, and the one, perceived, experienced and taken part in during its wakeful state, - are judged by it as being actual-looking, factual-looking, real-looking, genuine-looking or authentic-looking plus material-looking, substantial-looking, physical-looking, objective-looking or concrete-looking as well as manifold or diverse looking whenever it encounters them in their respective milieu or setting.

Human consciousness in the mode of dream sleep state judges the dream sleep world to be real and material even though it is made of dream stuff.

Human consciousness in the mode of wakeful state judges the wakeful state world to be real and material even though it is made of dream stuff.

From the perspective of human consciousness, which is in the mode of dream sleep state, only the dream sleep world and its people and consciousnesses exist and are real. The wakeful state world and its people and consciousnesses do not exist.

From the perspective of human consciousness, which is in the mode of wakeful state, the *dream sleep world* and its people plus consciousnesses are all made of dream stuff and therefore are neither real nor worth any further consideration.

So, the million-dollar question is "Who is right"?

Is the human consciousness in the mode of dream sleep state right or the human consciousness in the mode of the wakeful state right?

One must also bear in mind that human consciousness in the mode of dream sleep state and the human consciousness in the mode of wakeful state are not two different human consciousnesses.

They are two different avatars or incarnations of one and the same human consciousness.

The difference pertaining between these two different avatars or incarnations of one and the same human consciousness is that the one avatar or incarnation exists in two versions i.e., bodiless version on one hand and embodied version on the other, during its one and the same *AWARENESSBAL MODE* called the dream sleep state or the dream sleep mode, whereas its other avatar or incarnation exists in only one version i.e., embodied version during its other *AWARENESSBAL MODE* called the wakeful state or the wakeful mode.

The dream sleep state world and the wakeful state world, both generate the same emotional effect upon the respective or relevant avatar or incarnation of one and the same human consciousness when that avatar or incarnation is made to experience its respective or relevant world by the ubiquitous and infinite field of consciousness aka cosmic space aka god aka source or spring of human consciousness aka creator of the *physical matter.*

Cosmic space is the creator of both worlds i.e., dream sleep state world and the wakeful state world which are the subject of discussion as well as the object of comparison at this moment.

Therefore, it is not appropriate for human consciousness during its wakeful state, to judge the above two magical worlds, both of whom are DREAMAL WORLDS or *dream-stuff-composed-worlds,* unequally or differently. They both must be treated on a par with each other or equal to each other or coequal.

That is to say, they both must be accepted as being merely of magical nature, nothing more nothing less.

No such label or tag as "this magical world i.e., wakeful state world, is one hundred percent real" and "the other magical world i.e., dream sleep state world, is one hundred percent unreal" must be attached to either of them. To do so will be a big folly.

Both worlds are magical worlds and therefore, both are only partially real or temporarily real or mithyavically real and not absolutely real, permanently real or satyavically real.

That is to say, they both are real only as long as they last or endure, no more no less, because they both are transient as all magic is.

Both magical worlds i.e., dream sleep state world on one hand and the wakeful state world on the other, have been brought into being or, have been forged or fabricated in such a way by the ubiquitous and infinite field of consciousness aka cosmic space aka god aka source or spring of human consciousness aka creator of the *physical matter,* that they both will disappear completely at the moment of their death or demise or, at the moment of their final exit, without leaving any trace behind. This they will do by dissolving, merging, melding or, if it is preferred, by vanishing, disappearing or melting into the ubiquitous and infinite field of consciousness aka cosmic space aka god aka source or spring of human consciousness aka creator of *physical matter* who has created them both through the very ordinary or common-'o'-garden activity of daydreaming or oneiricking on its part and nothing else.

What has been said above can be put in another way.

Both magical worlds under discussion i.e., the dream sleep state world as well as the wakeful state world, which are encountered and therefore perceived, experienced and taken part in by the human consciousness, are transient, temporary or ephemeral in nature and therefore their reality is also of transient, temporary or ephemeral duration. That is to say, they both are real only for the duration they endure, last or exist and no more. They are not eternal or they are not eternally real or eternally existent in the manner the ubiquitous and infinite field of consciousness aka cosmic space aka god aka source or spring of human consciousness aka creator of *physical matter* is.

The ubiquitous and infinite field of consciousness aka cosmic space aka god aka source or spring of human consciousness aka creator of *physical matter* is the creator or maker of both magical worlds.

In other words, the ubiquitous and infinite field of consciousness aka cosmic space aka god aka source or spring of human consciousness aka creator of *physical matter* is the magician of even the magic world called the dream sleep state world which is encountered, perceived, experienced and taken part in by human consciousness during its dream sleep state, notwithstanding the cosmic delusion or the brahmandic vibhranti from which human consciousness suffers that the magical world called the dream sleep state world is its own creation and not of any outside agency.

The ubiquitous and infinite field of consciousness aka cosmic space aka god aka source or spring of human consciousness aka creator of *physical matter* inserts, shoves or thrusts the magical world or the magic show called the dream sleep state world into human consciousness during the latter's dream sleep state as per its own whim or fancy or, as per its own impulse or urge, in order to provide a clue to human consciousness that nothing is under its control, not even what happens to it during its dream sleep state and also to provide it an opportunity to fathom out the true or real nature of the magical world or the magic show it encounter, perceives as well as experiences and in which it takes part each day during its wakeful state.

To recapitulate.

Both magical worlds i.e., dream sleep state world and wakeful state world, which are encountered, perceived, experienced and taken part in by human consciousness 24/7 or around-the-clock have been created by the ubiquitous and infinite field of consciousness aka cosmic space aka god aka source or spring of human consciousness aka creator of *physical matter*. And the latter has also willed that human consciousness will encounter them plus perceive, experience and take part in them.

When one states that both worlds i.e., dream sleep state world and wakeful state world are magical in nature; it is not to say that both these worlds are fibs, falsehoods, fabrications or fictions or, both these worlds are untruths. It is to say instead, that, they both are as real as any magic show i.e., they are real as long as the magician performing the magic show continues with the magic. That's all, nothing more nothing less.

What has been said above can be put in another way.

When one states that the dream sleep state world and the wakeful state world are both magical in nature, one means that they both are being merely encountered, perceived, experienced and taken part in by the respective awatars or incarnations of human consciousness without the latter going very deeply into their underlying nature as to "how real they are"? Or "Are they both eternally real or only temporarily real"? That's all.

That statement certainly does not imply that they both are fibs, falsehoods, fabrications or fictions or, untruths.

The fact is that both these worlds are neither real nor unreal.

That is to say, they both are *MITHYAVIC WORLDS* i.e., they both are real as long as they exist, nothing more nothing less. They are not eternal in the manner their creator namely the ubiquitous and infinite field of consciousness aka cosmic space aka god aka source or spring of human consciousness aka creator of *physical matter* is.

When they exist, they both abide or, they both are extant or present or, if preferred, when they exist, they both are around or in existence inside the ubiquitous and infinite field of consciousness aka cosmic space aka their creator or god and when they don't, even then they both abide inside the ubiquitous and infinite field of consciousness aka cosmic space aka their creator or god but in potential, latent or dormant form only or, if it is preferred, in unrealised or unmanifested form only.

They both become actual, factual, hard or definite or, if one prefers, manifest inside the ubiquitous and infinite field of consciousness aka cosmic space aka god only when the latter wants them to do so.

Otherwise they both continue to abide in their unmanifested or unrealised form or, in their latent, dormant or potential form inside the ubiquitous and infinite field of consciousness aka cosmic space aka god, waiting to become actual, factual, definite, hard or manifest inside the latter whenever it wants them to become actual, factual, definite, hard or manifest inside it, through the activity of daydreaming or oneiricking on its part.

The truth about both magical worlds i.e., dream sleep state world on one hand & wakeful state world on the other, which are encountered, perceived, experienced and taken part in by human consciousness is that they both are dreamal or phantasmal in nature. That is to say, they both are made of daydream-stuff or oneiric-stuff of the ubiquitous and infinite field of consciousness aka cosmic space aka god aka source or spring of human consciousness aka creator of *physical matter.*

Both these magical worlds have been created or made manifest or, if it is preferred, made actual, factual, or definite by the ubiquitous and infinite field of consciousness aka cosmic space aka god aka source or spring of human consciousness aka creator of *physical matter,* inside itself through the very ordinary or "common-'o'-garden" activity called daydreaming, oneiricking or reverieing on its part and nothing else.

Demotion or downgrading of actual-looking, factual-looking, real-looking, genuine-looking or authentic-looking plus material-looking, substantial-looking, physical-looking, objective-looking, or concrete-looking as well as manifold or variegated-looking dream sleep state world, phenomenal world or experiential world which human consciousness encounters, perceives, experiences and takes part in during its dream sleep state to a substandard, below par, inferior, abysmal, appalling, dreadful, trashy, rubbishy, second-rate, third-rate, or even tenth-rate status and, elevation or promotion of the actual-looking, factual-looking, real-looking, genuine-looking or authentic-looking plus material-looking, substantial-looking, physical-looking, objective-looking, or concrete-looking as well as manifold or variegated-looking wakeful state world, phenomenal world or experiential world which human consciousness encounters, perceives, experiences and takes part in its wakeful state to a top quality, top grade, first class, first-rate, second to none, grade A, five star, superlative, excellent, outstanding, exceptional, marvellous, magnificent, splendid, matchless, peerless, unparalleled, top-notch, fantastic, fabulous, or stellar status by human consciousness without proper investigation on its part about the true nature of them both from the most fundamental, or, intrinsic, bottom-line, or absolute vantage point, standpoint, viewpoint, stance, position, perspective, or frame of reference is very unfortunate indeed.

Human consciousness encounters and therefore perceives, experiences and takes part in the actual-looking, factual-looking, real-looking, genuine-looking or authentic-looking plus material-looking, substantial-looking, physical-looking, objective-looking or concrete-looking as well as manifold or variegated-looking dream sleep state world, phenomenal world or experiential world of its dream sleep state with the aid of its five consciousnessbal senses which are made operational inside it during its dream sleep state by the ubiquitous and infinite field of consciousness aka cosmic space aka god aka source or spring of human consciousness aka creator of *physical matter*.

The fact is that the *DREAMAL WORLD* or the *FANTASMAL WORLD,* or, should one say, the actual-looking, factual-looking, real-looking, genuine-looking or authentic-looking plus material-looking, substantial-looking, physical-looking, objective-looking, or concrete-looking, as well as, manifold or variegated-looking dream sleep state world or dream-sleep-state-phenomenal-world or dream-sleep-state-experiential-world which is encountered and therefore perceived, experienced and taken part in by human consciousness during its dream sleep state is created by the ubiquitous and infinite field of consciousness aka cosmic space aka god aka source or spring of human consciousness, aka creator of *physical matter* inside human consciousness during its dream sleep state in order to forewarn, tip off, caution, alert, or apprise it, if it cares to be forewarned, tipped off, cautioned, alerted or apprised that the actual-looking, factual-looking, real-looking, genuine-looking or authentic-looking plus material-looking, substantial-looking, physical-looking, objective-looking, or concrete-looking, as well as, manifold or variegated-looking wakeful state world or wakeful-state-phenomenal-world or wakeful-state-experiential-world which human consciousness encounters, perceives, experiences and takes part in during its wakeful state is of the same intrinsic nature as the dream sleep state world is, or the dream-sleep-state-phenomenal-world or dream-sleep state-experiential-world is, the dream sleep world or dream sleep-state-phenomenal-world or dream-sleep-state experiential-world or, should one say, the *DREAMAL WORLD* or *FANTASMAL WORLD,* which human consciousness encounters, perceives, experiences and takes part in during its dream sleep state.

Both magical worlds or, absolutely to the point, both the *DREAMAL* or *PHANTASMAL WORLDS* i.e., dream sleep state world on one hand, and the wakeful state world on the other, have been created by cosmic space aka god aka source or spring of human consciousness aka creator of *physical matter* by the very ordinary or common-'o'-garden activity of daydreaming or oneiricking on its part and nothing else; notwithstanding the cosmic delusion or the brahmandic vibhranti which human consciousness harbours or holds or, nurses or cherishes or, clings or entertains which makes it passionately believe that the magical world or, absolutely to the point, *DREAMAL* or *PHANTASMAL WORLD* i.e., material, substantial, physical, objective or concrete world, it encounters, perceives and experiences as well as takes part in during its wakeful state is real while the other magical world or, absolutely to the point, *DREAMAL* or *PHANTASMAL WORLD,* which it encounters, perceives and experiences as well as takes part in during its dream sleep state is unreal. As said earlier, it is the fault of its cosmic delusion or brahmandic vibhranti from which it suffers, which makes it passionately believe that one magical world or, absolutely to the point, *DREAMAL* or *PHANTASMAL WORLD* i.e., wakeful state world amongst these two, is real while the other magical worlds or, absolutely to the point, *DREAMAL* or the *PHANTASMAL WORLDS* i.e., dream sleep state world is unreal. The true position with regards to them both is that they both are on a par with each other, or, they both are coequal.

What has been said above can be put in another way.

Both worlds, namely, the actual-looking, factual-looking, real-looking, genuine-looking or authentic-looking plus material-looking, substantial-looking, physical-looking, objective-looking, or concrete-looking, as well as, manifold or variegated-looking wakeful state world or wakeful state phenomenal-world or wakeful state experiential-world which is encountered, perceived, experienced and taken part in by the human consciousness during its wakeful state on one hand, and the actual-looking, factual-looking, real-looking, genuine-looking or authentic-looking plus material-looking, substantial-looking, physical-looking, objective-looking, or concrete-looking, as well as, manifold or variegated-looking dream sleep state world or dream-sleep-state-phenomenal-world or dream-sleep-state-experiential-world which is encountered , perceived, experienced and taken part in by the human consciousness during its dream sleep state on the other, have been created by cosmic space aka god aka source or spring of human consciousness aka creator of *physical matter* by the very ordinary or common-'o'-garden activity called daydreaming or oneiricking on its part and nothing else, notwithstanding the cosmic delusion or brahmandic vibhranti from which human consciousness suffers which makes the latter passionately believe that the *DREAMAL* or *PHANTASMAL COSMOS* i.e., physical-looking, material-looking, substantial-looking, objective-looking or concrete-looking cosmos it perceives, experiences and participates in during its wakeful state is more real than the *DREAMAL* or *PHANTASMAL COSMOS* i.e., physical-looking, material-looking, substantial-looking, objective-looking or concrete-looking cosmos it perceives, experiences and participates in during its dream sleep state.

As said earlier, it is the cosmic delusion or brahmandic vibhranti of human consciousness; cosmic delusion or brahmandic vibhranti which has been heaped upon or bequeathed to human consciousness by the ubiquitous and infinite field of consciousness aka cosmic space aka god aka source or spring of human consciousness aka creator of *physical matter;* which makes the human consciousness mindlessly, ignorantly, illiterately or unintelligently plus passionately believe that one amongst these two phenomenal-worlds or experiential-worlds is one hundred percent real while the other phenomenal-world or experiential-world is one hundred percent unreal i.e., that the wakeful state world or wakeful-state-phenomenal-world or wakeful-state-experiential-world, which it encounters, perceives, experiences and takes part in during its wakeful state is one hundred percent real while the dream sleep state world or dream-sleep-state-phenomenal-world or dream-sleep-state-experiential-world which it encounters , perceives, experiences and takes part in during its dream sleep state is one hundred percent unreal.

The factual position regarding both these phenomenal-worlds or experiential-worlds is that they both are on a par with each other or equal to each other or a match for each other or are coequal.

They both are only partially real or mithyavically real and not absolutely real or satyavically real in the manner, the ubiquitous and infinite field of consciousness aka cosmic space aka god aka source or spring of human consciousness aka creator of *physical matter* is absolutely real or satyavically real.

That is to say, neither of them is falsehood, unreal or asatya, but at the same time, neither of them is one hundred percent truth or absolute truth or satya either.

Instead, they are both partial truths, mithyas, relative truths or mithyavic truths.

Partial truths, mithyas, relative truths or mithyavic truths have many other names. For example :-

Delusional truths or, mayavic truths; magical truths or, mayajalic truths, dreamal truths or, swapnil truths etc.

Human consciousness during its dream sleep state, encounters and therefore, perceives, experiences and takes part in the *DREAMAL* or *PHANTASMAL WORLD* called the dream sleep state world with the aid of its five consciousnessbal senses and not with the aid of its five physical senses because during its dream sleep state, human consciousness is unaware of the existence of its physical body and along with the unawareness of its physical body during this dream sleep state, it is also unaware of the existence of the five physical senses of its physical body.

The five consciousnessbal senses, which are switched on, turned on, or become operational or activated inside human consciousness during its dream sleep state in order to enable or empower it to perceive, experience and take part in the dream sleep state world or, the *DREAMAL* or *PHANTASMAL WORLD* which it encounters during its dream sleep state, are nothing but the innate powers of human consciousness.

In other words, human consciousness itself, directly, perceives, experiences and takes part in the dream sleep state world or, in the *DREAMAL* or *FANTASMAL WORLD* which it encounters, faces, comes into contact with, or runs into, during its dream sleep state.

Thus, during its dream sleep state, human consciousness does not require the help of any outside agency such as physical senses of any form or shape of physical body i.e., *physical matter,* for perceiving, experiencing and taking part in the dream sleep state world or the *DREAMAL* or *FANTASMAL WORLD* which it encounters, faces, comes into contact with, or runs into, during its dream sleep state.

In contrast, the assistance of the physical senses i.e., *physical matter* is vital for human consciousness for perceiving, experiencing and taking part in the *DREAMAL* or *FANTASMAL WORLD* called the wakeful state world which it encounters, faces, comes into contact with, or runs into, during its wakeful state.

The assistance of *physical matter* composed physical senses is vital for human consciousness to perceive, experience & take part in the *DREAMAL* or *FANTASMAL WORLD* called the wakeful state world which it encounters, faces, comes into contact with, or runs into, during its wakeful state because the *DREAMAL* or *FANTASMAL WORLD* called the wakeful state world which it encounters, faces, comes into contact with, or runs into, during its wakeful state is not situated or placed or stationed inside human consciousness in the manner the dream sleep state world is, or, in the manner the *DREAMAL* or *FANTASMAL WORLD,* which human consciousness encounters, faces, comes into contact with, or runs into during its dream sleep state, is.

The *DREAMAL* or *FANTASMAL WORLD* or, the phenomenal or experiential world called the wakeful state world, which human consciousness encounters and therefore, perceives, experiences and participates in during its wakeful state, is situated, placed or stationed inside the consciousness of the daydreamer or oneiricker who has daydreamed or oneiricked this *DREAMAL* or *FANTASMAL WORLD* or, this phenomenal or experiential world called the wakeful state world.

And the daydreamer or oneiricker of this *DREAMAL* or *FANTASMAL WORLD* or, this phenomenal or experiential world called the wakeful state world is no less a being than the ubiquitous and infinite field of consciousness aka cosmic space aka god aka source or spring of human consciousness aka creator of *physical matter.*

What has been said above with regards to the underlying nature of the wakeful state world which is encountered, perceived, experienced and participated in, by the human consciousness during its wakeful state, can be paraphrased in the following manner.

The *DREAMAL* or *FANTASMAL WORLD* or, the phenomenal or experiential world called the wakeful state world which human consciousness encounters and therefore, perceives, experiences and participates in during its wakeful state is a fantastic or incredible daydream or oneiric of the ubiquitous and infinite field of consciousness aka cosmic space aka god aka source or spring of human consciousness aka creator of the *physical matter.*

From what has been said above about the underlying nature of the wakeful state world on one hand, and the underlying nature of cosmic space on the other, it transpires, flows, follows, emerges, or ensues that the physical body of human consciousness and all the other physical items which human consciousness encounters, perceives and experiences inside the wakeful state world during its wakeful state such as countless moons, planets, stars, black holes, galaxies and the like, are all composed of daydream stuff or oneiric stuff of cosmic space. And the latter, i.e., cosmic space is an incredible, ubiquitous and infinite field of consciousness which is the source or spring of human consciousness on one hand and the creator of the physical matter on the other. This incredible, ubiquitous and infinite field of consciousness aka cosmic space, which is the source or spring of human consciousness on one hand and the creator of the physical matter on the other, can be given whatever other names one fancies including god so long as one realises what it fundamentally or truly is and not what one believes or imagines it is.

To paraphrase.

From what has been said above about the underlying nature of the wakeful state world on one hand, <> (the wakeful state world which human consciousness encounters, perceives, experiences, and participates in during its wakeful state), <> and the underlying nature of cosmic space on the other, <*> (cosmic space which human consciousness also encounters during its wakeful state and visualises it, or, sees, notices, detects or discerns it, and additionally, about which it constantly wonders as to "*What is it* or, what is its underlying nature, and, whether the discovery of its underlying nature will answer the age old question as to the purpose of the *womb to tomb journey is vis-a-vis all living beings including man"?*) <*> it transpires or, follows, emerges, ensues or flows that the physical body of human consciousness and all the other physical items which human consciousness encounters, perceives and experiences inside the wakeful state world during its wakeful state such as countless moons, planets, stars, black holes, galaxies and the like, are all composed of daydream stuff or oneiric stuff of cosmic space. And the latter, i.e., cosmic space is an incredible, ubiquitous and infinite field of consciousness which is the source or spring of human consciousness on one hand and the creator of physical matter on the other. This incredible, ubiquitous and infinite field of consciousness aka cosmic space, which is the source or spring of human consciousness on one hand and the creator of physical matter on the other, can be given whatever other names one fancies including god so long as one realises what it fundamentally or truly is and not what one believes or imagines it is.

To restate.

What has been said above vis-a-vis the *fundamental nature* of the wakeful state world on one hand, <> (the wakeful state world which human consciousness encounters and therefore, perceives, experiences and participates in it only during its wakeful state), <> and the *fundamental nature* of cosmic space on the other; <*> (cosmic space, which too human consciousness encounters and therefore, sees or visualises only during its wakeful state and inside which, human consciousness discovers during its wakeful state, that its physical body is spatially situated or placed. Incidentally, the existence of the physical body too, around itself, human consciousness discovers only during its wakeful state and in its no other state, for example, dream sleep state); <*> it transpires or, follows, emerges ensues or flows or, comes to light that the physical body of human consciousness as well as all the other physical items which human consciousness encounters and therefore perceives and experiences inside the wakeful state world during its wakeful state such as moons, planets, stars, black holes, galaxies and the like, are all composed of daydream stuff or oneiric stuff of cosmic space. And the latter, i.e., cosmic space is an incredible, ubiquitous and infinite field of consciousness which is the source or spring of human consciousness on one hand and the creator of physical matter on the other.

In other words, cosmic space is the author, creator or progenitor of physical-looking, material-looking, substantial-looking, objective-looking or concrete-looking *cosmos, universe, or world* which human consciousness encounters and therefore perceives, experiences and participates in during its wakeful state as a caged or captive, or, as a bound or bounded, or, as a circumscribed, confined, defined, distinct, delineated or silhouetted, or, as an individual consciousness.

The caged or captive, bound or bounded, circumscribed, confined, defined, distinct, delineated or silhouetted, or, individual human consciousness has become circumscribed, delineated, silhouetted, separated, segregated, sequestered, cut off or isolated from its source or spring namely the surrounding, ubiquitous and infinite ocean of consciousness aka cosmic space aka god aka source or spring of human consciousness aka creator of *physical matter* due to the presence of a thick layer of physical-looking, substantial-looking, objective-looking or concrete-looking *matter* around it namely the physical-looking, material-looking, substantial-looking, objective-looking or concrete-looking *body.*

To rehash.

From what has been said above about the nature of the wakeful state world on one hand, <> (wakeful state world which human consciousness encounters and therefore, perceives, experiences and takes part in during its wakeful state), <> and about the nature of cosmic space, on the other, <*> (cosmic space which is present only in the wakeful state world and therefore, is encountered plus visualised, seen, discerned or noticed by human consciousness only during its wakeful state and during no other state, for example, during its dream sleep state and inside which it is discovered by human consciousness that its physical body is spatially placed or situated, its physical body whose presence around itself, human consciousness also discovers only during its wakeful state and during its no other state, for example, during its dream sleep state); <*> it transpires or, follows, emerges, ensues or flows that the physical body of human consciousness ~*~ (whose presence around itself human consciousness discovers only during its wakeful state and during no other state, for example, during its dream sleep state) ~*~ is composed of daydream stuff or oneiric stuff of cosmic space. And the latter i.e., cosmic space is the source or spring of human consciousness on one hand and the creator of physical matter on the other.

In other words, cosmic space is the creator of physical-looking, material-looking, substantial-looking, objective-looking or concrete-looking *cosmos, universe, or world* which human consciousness encounters perceives, experiences and participates in during its wakeful state as a caged or captive, bound or bounded, circumscribed, confined, defined, distinct, delineated or silhouetted, or, individual consciousness. Human consciousness has become circumscribed, delineated, silhouetted, separated, segregated, sequestered, cut off or isolated from the surrounding, ubiquitous and infinite ocean of consciousness aka cosmic space aka source or spring of human consciousness aka creator of *physical matter,* due to the presence of a thick layer of physical-looking, substantial-looking, objective-looking or concrete-looking *matter* around it namely the physical-looking, material-looking, substantial-looking, objective-looking or concrete-looking *body.*

Another conclusion that needs to be drawn from all that has been said above is the following :-

Cosmic space, which is seen by human consciousness via the media of the eyes of its physical body during its wakeful state <> (and not during its dream sleep state) <> is an extraordinary, ubiquitous and infinite field of consciousness of one and only God or one and only Brahman aka The source or spring of human consciousness aka The Creator of physical-looking, substantial-looking, objective-looking or concrete-looking *matter* i.e., The Creator of physical-looking, material-looking, substantial-looking, objective-looking or concrete-looking *cosmos, universe, or world* which human consciousness encounters and therefore perceives, experiences and participates in during its wakeful state as a caged or captive, bound or bounded, or, as a confined, defined, distinct, delineated or silhouetted, or as a circumscribed or, as an individual consciousness.

One further conclusion also needs to be arrived at from

all that has been said above. This conclusion is the following :-

Physical-looking body of human consciousness <> (physical-looking body of human consciousness whose presence human consciousness is aware as an appendage, auxiliary, accessory, adjunct, add-on, addition or attachment around itself only during its wakeful state and not during its dream sleep state) <> and the rest of the physical-looking entities or bodies of the physical-looking world, cosmos or universe <*> (whose presence too human consciousness is aware only during its wakeful state and not during its dream sleep state) <*> are all sited, situated, located or, all stationed or positioned inside the *consciousness* or *mind* of the daydreamer or oneiricker who has oneiricked the physical-looking matter and therefore physical-looking world, cosmos or universe, physical-looking world, cosmos or universe which is encountered and therefore perceived and experienced plus handled and interacted with by human consciousness only during its wakeful state and not during its dream sleep state. This *consciousness* or *mind* of the daydreamer or oneiricker who has oneiricked the physical-looking matter and therefore physical-looking world, cosmos or universe is none other than the entity called cosmic space.

To recapitulate.

From what has been said above, ~<*>~ with regards to the daydreamal or oneirical nature or, with regards to the daydream-stuff-composed or oneiric-stuff-composed nature of the physical-looking world, cosmos or universe which human consciousness encounters and therefore perceives, experiences and takes part in during its wakeful state, and with regards to the ubiquitous and infinite field of consciousness aka cosmic space aka god aka source or spring of human consciousness aka creator of the physical-looking matter and therefore, creator of the physical-looking world, cosmos or universe, being the daydreamer or oneiricker of the daydream-stuff-composed or oneiric-stuff-composed world, cosmos or universe, which human consciousness encounters and therefore perceives, experiences and takes part in during its wakeful state, ~<*>~ it flows, follows, emerges or ensues or, it transpires that the physical-looking body and therefore physical-looking senses of human consciousness, whose existence human consciousness is aware only during its wakeful state and not during its dream sleep state, are both composed of daydream stuff or oneiric stuff of the ubiquitous and the infinite field of consciousness aka cosmic space aka god aka source or spring of human consciousness aka creator of the physical-looking matter and therefore, creator of the physical-looking world, cosmos or universe which human consciousness encounters, perceives, experiences and takes part in during its wakeful state.

One further conclusion flows, follows, emerges or ensues from what has been said above. This conclusion is the following :-

The Physical-looking body of human consciousness <> (physical-looking body of human consciousness whose presence human consciousness is aware of, around itself as an appendage, auxiliary, accessory, adjunct, add-on, addition or attachment only during its wakeful state and not during its dream sleep state) <> and the rest of the physical-looking entities or bodies of the physical-looking world, cosmos or universe <*> (whose presence too human consciousness is aware only during its wakeful state and not during its dream sleep state) <*> are all sited, situated or located or, all placed, positioned or stationed inside the *consciousness* or *mind* of the daydreamer or oneiricker who has daydreamed or oneiricked all the physical-looking entities or bodies of the physical-looking world, cosmos or universe which are encountered and therefore, perceived, experienced plus handled and interacted with by human consciousness during its wakeful state.

And this *consciousness* or *mind* of the daydreamer or oneiricker of the physical-looking world, cosmos or universe is none other than the entity called cosmic space.

Therefore, cosmic space which is seen by the human consciousness via the media of its physical eyes during its wakeful state <> (and not during its dream sleep state) <> is not an unremarkable, humdrum, mundane, colourless, dull, boring, lacklustre, uninteresting or uninspiring thing. Neither is it an existence, being, truth or reality which lacks or is devoid of unusual or special mystery, enigma, conundrum, secret, riddle or puzzle buried deep inside it.

On the contrary, cosmic space, which is seen by the human consciousness with the help of its physical eyes during its wakeful state is the consciousness of god or the mind of god. This consciousness of god or mind of god aka cosmic space is the creator of the physical-looking matter and therefore, creator of the physical-looking world, cosmos or universe which human consciousness encounters, perceives, experiences and takes part in during its wakeful state. Inside this consciousness or mind of god aka cosmic space, the physical-looking world, cosmos or universe, consisting of countless physical-looking moons, planets, stars, black holes & galaxies, is floating, wafting, or levitating, plus whirling, twirling or spiraling non-stop and has been doing so from the beginning of the current time which was some 13.7 billion light years ago and will continue to do so till the end of the current time.

The physical-looking body of human consciousness, whose presence around itself human consciousness becomes aware of, only during its wakeful state, constitutes only an infinitesimally small part of the entire physical-looking world, cosmos, or universe. Human consciousness is not aware of the presence of the physical-looking body around itself during its dream sleep state. The existence of physical-looking cosmos too, human consciousness becomes aware of only during its wakeful state. It is not aware of the existence of the physical-looking cosmos during its dream sleep state.

The physical-looking body of human consciousness, whose presence around itself human consciousness becomes aware of only during its wakeful state; it is a mere appendage, auxiliary, accessory, adjunct, add-on, addition or attachment, or, absolutely to the point, is a mere circumscribing, limiting, confining or, restraining obstacle or wall or, a separator, dam or dyke which circumscribes, limits, confines or restrains it from merging or coalescing into or, uniting or becoming one with its source or spring, namely, the surrounding, ubiquitous and infinite ocean of consciousness aka cosmic space aka god aka creator of the physical-looking matter and therefore, creator of the physical-looking world, cosmos or the universe which human consciousness encounters, perceives, experiences and takes part in during its wakeful state.

The physical-looking matter - which composes the physical-looking body of human consciousnesses on one hand and the remaining bodies of the physical-looking world, cosmos or universe on the other - has been availed, utilised, made use of, brought into play or put into service by cosmic space aka god aka source or spring of human consciousness aka creator of the physical-looking matter, as an agent, tool, instrument or implement to create variety, diversity, heterogeneity or multiplicity inside its nondescript or featureless, expanded, distended, dilated or expanded consciousness during its activity of daydreaming or oneiricking so as to amuse, entertain or regale itself and nothing else.

The term "consciousnessbal senses" one has employed in the caption of this chapter, as a metaphor or, figure of speech to communicate the quintessential truth that consciousness does not necessarily require the help of physical-looking senses to perceive, experience and take part in a *PHENOMENAL or EXPERIENTIAL world* if the *PHENOMENAL or EXPERIENTIAL world* in question is situated inside that consciousness itself as is the case during the dream sleep state of human consciousness as well as during the daydreaming or oneiricking state of human consciousness. That is to say, when human consciousness is in its dream sleep state as well as when human consciousness daydreams or oneirics in its wakeful state.

During the dream sleep state of human consciousness, the latter's dream sleep world exists inside it only and nowhere else. Therefore, human consciousness during its dream sleep state, does not require the help of physical-looking senses to perceive, experience and take part in its dream sleep world.

Similar is the situation when human consciousness daydreams or oneirics while awake. In the latter instance too, daydream or oneiric, created by human consciousness in its wakeful state is situated inside human consciousness itself and nowhere else.

Therefore, to perceive, experience and take part in its daydream world or oneiric world, human consciousness does not require the help of physical-looking senses of its physical-looking body, physical-looking body, whose presence it feels around itself as an add-on, appendage, accessory or attachment, or, better still, as an albatross, burden, constraint, cross to bear, encumbrance, hindrance, liability, load, millstone, nuisance, strain, trouble, worry or weight, or, absolutely to the point, as a circumscribing, limiting, confining or restraining obstacle or wall or, a separator, dam or dyke which is circumscribing, limiting, confining or restraining it from merging or coalescing into or, uniting or becoming one with its source or spring, namely, the surrounding, ubiquitous and infinite ocean of consciousness aka cosmic space aka god aka creator of physical-looking matter aka creator of physical-looking world, cosmos or universe.

As said earlier, human consciousness perceives, experiences and takes part in the *DREAMAL WORLD* or the *FANTASMAL WORLD* of its dream sleep state, or, if it is preferred, human consciousness perceives, experiences and takes part in the *DREAM SLEEP WORLD* of its dream sleep state with the aid of its consciousnessbal senses and not with the aid of the physical-looking senses of its physical-looking body as it does during its wakeful state in order to perceive, experience and take part in the *DREAMAL WORLD* or *FANTASMAL WORLD* of its wakeful state, that is to say, in order to perceive, experience and take part in the physical-looking or material-looking world of its wakeful state.

To paraphrase.

Human consciousness during its wakeful state, uses the physical-looking senses of its physical-looking body in order to perceive, experience and take part in the physical-looking *DREAMAL WORLD* or *FANTASMAL WORLD* of its wakeful state which is situated external to it or outside it and not within it or inside it as is the case with regards to the dream sleep state world which is situated within it or inside it. The physical-looking *DREAMAL WORLD* or *FANTASMAL WORLD* which human consciousness encounters, perceives, experiences and participates in during its wakeful state and which is situated external to it or outside it and not within it or inside it (as is the case with regards to the dream sleep state world), is also called the material-looking, substantial-looking, objective-looking or concrete-looking world, encountered, perceived, experienced and participated in by human consciousness during its wakeful state.

As said before, human consciousness, in contrast, during its dream sleep state uses its consciousnessbal senses to perceive, experience and take part in the physical-looking *DREAMAL* or *FANTASMAL WORLD* or the physical-looking *DREAM SLEEP STATE WORLD* of its dream sleep state.

To recapitulate.

Human consciousness uses the physical-looking senses of its physical-looking body during its wakeful state only, in order to perceive, experience and take part in the physical-looking world which it encounters during its wakeful state.

Human consciousness does not use the physical-looking senses of its physical-looking body during its dream sleep state for the purpose of perceiving, experiencing and taking part in the dream sleep state world which it encounters during its dream sleep state because it loses the awareness of the existence of its physical-looking body and therefore the awareness of the existence of the latter's physical-looking senses too during its dream sleep state.

One's purpose behind bringing to the notice of human consciousness the fact that physical senses are essential for human consciousness only when it wants to use them to perceive, experience and take part in a phenomenal world or experiential world which is situated outside or external to human consciousness and not when that phenomenal world or experiential world is situated inside the human consciousness.

Since the phenomenal world or experiential world which human consciousness encounters, perceives, experiences and takes part in, during its wakeful state, is situated outside or external to human consciousness, the possession on the human consciousness's part of a physical body and body's physical senses and the rest, becomes absolutely vital for it because without the possession on its part of a physical body and the latter's physical senses and the rest, even the existence of human consciousness as a separate, unattached, or independent creature or being or, truth or reality is not possible, a creature or being or, truth or reality who is apart, autonomous, or distinct from the ubiquitous and boundless ocean of consciousness called cosmic space aka god aka source or spring of human consciousness aka creator of physical matter aka creator of physical cosmos.

What has been said above can be put in another way.

Since the phenomenal world or experiential world which human consciousness encounters, perceives, experiences and takes part in during its wakeful state is situated outside or external to human consciousness, the possession on its part of a physical body and the latter's physical senses and the rest becomes absolutely vital for it because without them even the existence of human consciousness as a free-standing or self-contained consciousness who is cut off or fenced off from or, who is isolated, segregated or disjoined from the surrounding, ubiquitous and infinite ocean of consciousness called cosmic space aka god aka source or spring of human consciousness aka creator of physical matter aka creator of physical cosmos, is not possible

COSMIC SPACE PERCEIVES & GETS A KICK OUT OF ITS CONSCIOUSNESSBAL IMAGERY, DREAMRY, DAYDREAM OR ONEIRIC, UTILISING ITS CONSCIOUSNESSBAL SENSES.

Let ne reiterate some paramount points before one deals with the main theme of this chapter.

During its wakeful state, what human consciousness unenlightenedly dubs, calls, or tags as physical cosmos, in reality, is nothing of the sort.

Instead, it is a mere observational and experiential imagery, dreamry, daydream or oneiric of cosmic space or Brahmandic Aakash and nothing else which it is observing, beholding or perceiving by utilising its *consciousnessbal senses.* Cosmic space or Brahmandic Aakash gets a real kick or joy out of observing, beholding or perceiving this imagery, dreamry, daydream or oneiric of its. The purpose of cosmic space or Brahmandic Aakash behind creation of this imagery, dreamry, daydream or oneiric of its, is to derive fun, amusement or entertainment out of observing, beholding or perceiving it.

Thus, inside this cosmic space or Brahmandic Aakash, all the moons, planets, stars, black holes, galaxies and the like, (which are unenlightenedly imagined by human consciousness as being of physical nature), are floating, levitating, or wafting plus whirling, twirling or spiraling non-stop as a mere, perceptual and experiential, or, as a mere observational and enjoyational imagery, dreamry, daydream or oneiric of this cosmic space or Brahmandic Aakash, and nothing else.

These moons, planets, stars, black holes, galaxies and the like have been floating, levitating, or wafting plus whirling, twirling or spiraling non-stop as a mere, perceptual and experiential, or, as a mere observational and enjoyational imagery, dreamry, daydream or oneiric of this cosmic space or Brahmandic Aakash and nothing else, from the beginning of the current cosmic time and will continue to do so till the end of the current cosmic time.

The imagery, dreamry, daydream or oneiric of cosmic space or Brahmandic Aakash, which is unenlightenedly imagined as being made or composed of physical matter by human consciousness in its wakeful state, in reality, is made or composed of a congealed, condensed, concentrated, coagulated, clotted, compacted or compressed section, segment, part or portion of awareness, consciousness, sentience or mind of cosmic space or Brahmandic Aakash and nothing else.

The section, segment, part or portion of awareness, consciousness, sentience or mind of cosmic space or Brahmandic Aakash (which has been utilized by the latter to form or make physical matter out of itself), has been congealed, condensed, concentrated, coagulated, clotted, compacted or compressed by cosmic space or Brahmandic Aakash through a very ordinary or common-'o'-garden activity on its part called the activity of daydreaming, oneiricking or reverieing or, if it is preferred, awarenessbal, consciousnessbal, sentiencel or mental imagery-making or dreamry-making.

Therefore, the imagery, dreamry, daydream or oneiric of cosmic space or Brahmandic Aakash, which is unenlightenedly called the physical, material, substantial, objective or concrete cosmos, universe, or world by human consciousness in its wakeful state, is awarenessbal, consciousnessbal, mental or sentiencel in quintessence, core, heart, or centre or, if it is preferred, is awarenessbal, consciousnessbal, mental or sentiencel in nature, nub, nitty-gritty or crux because it is made or composed of a section, segment, part or portion of awareness, consciousness, sentience or mind of cosmic space or Brahmandic Aakash, albeit a congealed, condensed, concentrated, coagulated, clotted, compacted or compressed section, segment, part or portion of awareness, consciousness, sentience or mind of cosmic space or Brahmandic Aakash.

In other words, cosmic space or Brahmandic Aakash, which the human consciousness, during its wakeful state, sees or visualizes through the medium of eyes of the physical body, is the *imagery-maker, dreamry-maker* or *reverie-maker,* or, if it is preferred, is the *daydreamer, or oneiricker* of the awarenessbal, consciousnessbal, sentiencel or mental *imagery, reverie, dreamry, daydream, or oneiric* which is unenlightenedly imagined and therefore, labeled or tagged as being the physical cosmos by human consciousness during its wakeful state.

The unenlightened-ness of human consciousness vis-a-vis the actual or real nature of physical cosmos is on account of the human consciousness's total ignorance of the consciousnessbal cosmology on one hand and its *interest and involvement* purely in the physical cosmology, on the other.

The *unidirectional interest and involvement* of the human consciousness purely in the physical cosmology on one hand and the total lack of knowledge on its part of the consciousnessbal cosmology on the other, has led to its appalling or lamentable inability or failure to discover the true nature of cosmic space or Brahmandic Aakash plus human consciousness on one hand and the true nature of physical matter on the other.

From what has been said above, it transpires that cosmic space or Brahmandic Aakash is an incredible, ubiquitous and infinite field of consciousness of God or Brahman aka source or spring of human consciousness aka creator of *physical matter.*

Cosmic space or Brahmandic Aakash is the entity which human consciousness sees or visualises during its wakeful state through the medium of eyes of its physical body and inside which the perceptual and experiential or, the observational and enjoyational imagery, dreamry, daydream or oneiric of cosmic space or Brahmandic

Aakash, which is unenlightenedly called physical cosmos by human consciousness in its wakeful state, is territorially or spatially situated, sited, stationed or positioned.

Furthermore, cosmic space or Brahmandic Aakash i.e., the incredible, ubiquitous and infinite field of consciousness of God or Brahman aka source or spring of human consciousness aka creator of *physical matter,* which has created inside itself or, has given birth inside itself to its consciousnessbal imagery, dreamry, daydream or oneiric called the physical world, by a very ordinary or common-'o'-garden activity on its part called the consciousnessbal imagery-making, dreamry-making, daydreaming, or oneiricking, merely to amuse, entertain or regale itself and nothing else.

That is to say, cosmic space or Brahmandic Aakash viz., the incredible, ubiquitous and infinite field of consciousness of God or Brahman aka the source or spring of human consciousness aka the creator of *physical matter,* has given birth to its consciousnessbal imagery, dreamry, daydream or oneiric called the physical world, inside its consciousness called cosmic space or Brahmandic Aakash, by a very ordinary or common-'o'-garden activity on its part called the consciousnessbal imagery-making, dreamry-making, daydreaming, or oneiricking, merely to amuse, entertain or regale itself and nothing else.

Cosmic space or Brahmandic Aakash, as said before, is not an unremarkable, humdrum, mundane, colourless, dull, boring, lacklustre, uninteresting or uninspiring thing. Neither is it an existence, being, truth or reality which lacks or is devoid of unusual or special mystery, enigma, conundrum, secret, riddle or puzzle buried deep inside it.

On the contrary, cosmic space, which is seen by the human consciousness with the help of the eyes of its physical body during its wakeful state is the consciousness of god or the mind of god. This consciousness of god or the mind of god aka cosmic space is the creator of the physical-looking matter and therefore, creator of the physical-looking world, cosmos or universe which human consciousness encounters, perceives, experiences and takes part in during its wakeful state.

Inside this consciousness or mind of god aka cosmic space, the physical-looking world, cosmos or universe; consisting of countless physical-looking moons, planets, stars, black holes and galaxies, is floating, wafting, or levitating, plus whirling, twirling or spiraling non-stop and has been doing so from the beginning of the current cosmic time which was some 13.7 billion light years ago and will continue to do so till the end of the current cosmic time.

Let one explain the fundamental nature of cosmic space or Brahmandic Aakash further.

Cosmic space or Brahmandic Aakash is the 3-D or three-dimensional form or version of the dimensionless form or version of God or Brahman aka source or spring of human consciousness aka creator, maker or progenitor of *physical, substantial, objective* or *concrete matter.*

The dimensionless form or version of God or Brahman aka source or spring of human consciousness aka creator of *physical matter,* has transmuted, transformed or metamorphosed itself into its 3-D or three-dimensional form or version called cosmic space or Brahmandic Aakash by the expansion, distension, dilation or inflation of its dimensionless form or version.

The dimensionless form or version of God or Brahman aka source or spring of human consciousness aka creator of *physical matter* is the original, primal, primeval, primordial, autochthonous or autochthonic form or version of God or Brahman aka source or spring of human consciousness aka creator of *physical matter.*

The dimensionless form or version of God or Brahman

aka source or spring of human consciousness aka creator of *physical matter,* that is to say, the original, primal, primeval, primordial, autochthonous or autochthonic form or version of God or Brahman aka source or spring of human consciousness aka creator of *physical matter,* exists before the beginning of time and after the end of time.

After the beginning of time and before the end of time, the dimensionless form or version of God or Brahman aka source or spring of human consciousness aka creator of *physical matter,* that is to say, the original, primal, primeval, primordial, autochthonous or autochthonic form or version of God or Brahman aka source or spring of human consciousness aka creator of *physical matter* exists in its 3-D or three-dimensional form or version called cosmic space or Brahmandic Aakash which is nothing but the expanded, distended, dilated or inflated form or version of the original, primal, primeval, primordial, autochthonous or autochthonic form or version of God or Brahman aka source or spring of human consciousness aka creator of *physical matter.*

The 3-D or three-dimensional form or version of the dimensionless form or version of God or Brahman aka source or spring of human consciousness aka creator of *physical matter* has been given rise to by the process of expansion, distension, dilation or inflation of the dimensionless form or version of God or Brahman aka source or spring of human consciousness aka creator of *physical matter.*

That is to say, the 3-D or three-dimensional form or version of the dimensionless form or version of God or Brahman aka source or spring of human consciousness aka creator of *physical matter* has been given rise to or has been brought into being by the process of expansion, distension, dilation or inflation of the dimensionless form or version or, the original, primal, primeval, primordial, autochthonous or autochthonic form or version of God or Brahman aka source or spring of human consciousness aka creator of *physical matter,* that existed before the beginning of time and after the end of time.

The dimensionless form or version of God or Brahman aka source or spring of human consciousness aka creator of *physical matter,* that is to say, the original, primal, primeval, primordial, autochthonous or autochthonic form or version of God or Brahman aka source or spring of human consciousness aka creator of *physical matter,* which existed before the beginning of time and will again exist after the end of time, has transmuted, transformed or metamorphosed itself into its current 3-D or three-dimensional form or version, namely, the incredible, ubiquitous and infinite field of consciousness called cosmic space or Brahmandic Aakash in order to spatially accommodate its current 3-D or three-dimensional daydream or oneiric which is ignorantly called physical cosmos by human consciousness on account of its unenlightened-ness with regards to the true nature of the physical cosmos.

Having reiterated the paramount points relating to the subject-matter of this chapter, let one now deal with the subject-matter of this chapter itself.

Cosmic space or Brahmandic Aakash, <> (aka the incredible, ubiquitous and infinite, 3-D or three-dimensional field of consciousness of God or Brahman, aka the source or spring of human consciousness, aka the creator of congealed, condensed, concentrated, coagulated, clotted, compacted or compressed consciousnessbal-stuff, which is unenlightenedly or nesciently called or dubbed as being the *physical matter* of this cosmos by the human consciousness), <> is the consciousnessbal imagery-maker, dreamry-maker, reverie-maker, daydreamer, or oneiricker of its consciousnessbal imagery, dreamry, reverie, daydream or oneiric which is called, dubbed or tagged as being the *physical world, cosmos* or universe by the human consciousness on account of its nescience or unenlightened-ness of the actual, real, or true nature of the latter which, in turn, is due to the lack of *curiosity* and, of *knowledge* on its part, of the consciousnessbal cosmology plus due to the *unidirectional curiosity* and *unidirectional knowledge* on its part of the physical cosmology only.

As said above, cosmic space or Brahmandic Aakash is the incredible, ubiquitous and infinite, 3-D or three-dimensional field of consciousness of God or Brahman of the universe. It thus, also is the source or spring of human consciousness as well as the creator or maker of the congealed, condensed, concentrated, coagulated, clotted, compacted or compressed consciousnessbal-stuff, which is unenlightenedly or nesciently dubbed or tagged as being the *physical matter* of the cosmos by the human consciousness.

Cosmic space or Brahmandic Aakash, after creating or making its current consciousnessbal imagery, dreamry, reverie, daydream or oneiric; <> (which is called or dubbed as being the *physical world* by the human consciousness on account of its unenlightened-ness vis-a-vis the true nature of the latter); <> is now seeing, observing, watching, beholding or perceiving the same, constantly or non-stop with the help of its *consciousnessbal senses* in order to derive personal fun, amusement, entertainment or regalement.

That is to say, cosmic space or Brahmandic Aakash, after creating or making its current consciousnessbal imagery, dreamry, reverie, daydream or oneiric; <> (which, as said before, is called or dubbed as being the *physical world* by the human consciousness on account of if unenlightenedness vis-a-vis the true nature of the latter); <> is now constantly or non-stop observing, beholding or perceiving it with the help of its *consciousnessbal senses* in order to derive personal fun, amusement, entertainment or regalement.

The expression "*consciousnessbal senses*" is a metaphor or figure of speech which has been coined, conceived, devised, or has been dreamed up with the intent to draw the attention of human consciousness towards an innate power of all consciousnesses; irrespective of whether the consciousness in question is that of cosmic space aka Brahmandic Aakash i.e., God aka Brahman of the current universe, or, that of a human being i.e., human consciousness.

This innate power of all consciousnesses enables them to make or create inside themselves their own, personal or private consciousnessbal imagery, dreamry, reverie, daydream or oneiric through the instrumentality of a very ordinary or common-'o'-garden activity on their part called the consciousnessbal imagery-making, dreamry-making, reverie-making, daydreaming, oneiricking or reverieing. This innate power of their's also enables them to visualize, see, observe or perceive the latter, namely, their consciousnessbal imagery, dreamry, reverie, daydream or oneiric, with the aid of their inherent ability of visualization, observation or perception, in order to derive personal fun, joy, amusement, entertainment or regalement.

Cosmic space or Brahmandic Aakash i.e., God or Brahman of the current universe, at present is doing exactly the same as described above.

That is to say, cosmic space or Brahmandic Aakash i.e., God or Brahman of the current universe, at present has made or created its own, personal or private consciousnessbal imagery, dreamry, reverie, daydream or oneiric inside itself, with the aid of its own or innate power of consciousnessbal imagery-making, dreamry-making, daydreaming, or oneiricking.

It is another matter that, on account of cosmic space's or Brahmandic Aakash's conferred or bestowed *cosmic delusion* or brahmandic vibhranti upon it, human consciousness unenlightenedly or nesciently calls or dubs this consciousnessbal imagery, dreamry, reverie, daydream or oneiric of cosmic space or Brahmandic Aakash, as being of the *physical, material, substantial, or concrete* nature and not of the *dreamal, phantasmal* or *oneirickle* nature.

After making or creating its current consciousnessbal imagery, dreamry, reverie, daydream or oneiric inside itself, with the aid of its innate or inherent power of consciousnessbal imagery-making, dreamry-making, reverie-making, daydreaming, oneiricking or reverieing, *cosmic space or Brahmandic Aakash* is now seeing, observing, watching, beholding or perceiving it with the aid of its innate or inherent, *consciousnessbal power* of seeing, observing, watching, beholding or perceiving it. It is deriving real kick or joy out of seeing, observing, watching, beholding or perceiving this imagery, dreamry, daydream or oneiric of its. This imagery, dreamry, daydream or oneiric of its, is an extremely precious fruit of its immense consciousnessbal, awarenessbal, sentiencel, or mental labour.

The sole purpose of cosmic space or Brahmandic Aakash aka God or Brahman, behind the creation of this imagery, dreamry, daydream or oneiric is to derive personal fun, joy, amusement, entertainment or regalement out of seeing, observing, watching, beholding or perceiving it.

WHY COSMIC SPACE OR BRAHMANDIC AAKASH CANNOT TALK TO HUMAN CONSCIOUSNESS IN THE WAY HUMAN CONSCIOUSNESS TALKS TO OTHER HUMAN CONSCIOUSNESSES

Human consciousness wonders, or, is curious or fascinated to know that, if cosmic space or Brahmandic Aakash, <> (which it sees or visualizes with the help of the eyes of its physical body during its wakeful state and inside which the physical-looking world, cosmos or universe - consisting of countless physical-looking moons, planets, stars, black holes and galaxies - is floating, wafting, or levitating, plus whirling, twirling or spiraling non-stop and has been doing so from the beginning of the current cosmic time and will continue to do so till the end of the current cosmic time), <> is neither a boring, colorless, dull, humdrum, mundane, lackluster or uninteresting thing nor is it a truth, reality, existence, or, being, which lacks or is devoid of unusual or special mystery, enigma, conundrum, riddle, puzzle or secret, buried deep inside it, and instead or alternatively plus in utter, absolute or stark contrast, is the consciousness of god or Brahman, or, the mind of god or Brahman, and, this consciousness of god or Brahman or, the mind of god or Brahman is the creator or maker of the physical-looking matter and therefore, in turn, is the creator or maker of the physical-looking world, cosmos, or, universe, which the human consciousness encounters and therefore perceives, experiences and takes part in during its wakeful state, then why this cosmic space or Brahmandic Aakash aka the consciousness of god or Brahman aka the mind of god or Brahman aka the creator or maker of the physical-looking world, cosmos or universe, does not speak to human consciousness in the physical-looking style, fashion, or, manner, that is to say, in the style, fashion, or, manner, one human consciousness speaks to another human consciousness in its wakeful state.

The answer to the above question, which has been posed by the human consciousness, must be provided because it is totally logical, valid, well founded, or proper.

This answer is the following :-

Human consciousness is an embodied consciousness, whereas, in utter, absolute, or, stark contrast, cosmic space or Brahmandic Aakash aka the consciousness of god or Brahman aka the mind of god or Brahman, is unique or the only one of its kind and, therefore, an incredible or extraordinary, bodiless consciousness.

When one human consciousness talks to another human consciousness, it uses the speech-equipment of its physical-looking body as a tool to accomplish this feat or exercise.

This speech-equipment consists of body's lips and lungs plus larynx and tongue. It has been possible for human consciousness to have a speech-equipment of a physical-looking kind, only because of the possession on its part of a physical-looking body. The physical-looking body with its physical-looking *speech equipment,* consisting of lips and lungs plus larynx and tongue; has been bestowed or conferred on the human consciousness by none other than cosmic space or Brahmandic Aakash who is the consciousness of god or Brahman, or, who is the mind of god or Brahman aka the source or spring of human consciousness on one hand and the creator or maker of the physical-looking matter of the physical-looking cosmos, on the other.

It is this physical-looking matter of the physical-looking cosmos, which has been made use of, or, has been deployed, used, utilized, or employed by cosmic space or Brahmandic Aakash aka the consciousness of god or Brahman aka the mind of god or Brahman, in order to make, form or create, the physical-looking body of human consciousness plus this body's physical-looking *speech-equipment* and the rest.

Since cosmic space or Brahmandic Aakash aka the consciousness of god or Brahman aka the mind of god or Brahman aka the creator of the physical-looking world, is uniquely and incredibly bodiless, that is to say, since cosmic space or Brahmandic Aakash aka the consciousness of god or Brahman aka the mind of god or Brahman aka the creator of the physical-looking world, uniquely and incredibly does not *live, exist, abide,* or *subsist* in physical-looking form in the manner human consciousness does, and instead, *lives, exists, abides,* or *subsists* in bodiless form, it lacks or is devoid of physical-looking *speech-equipment,* consisting of lips and lungs plus larynx and tongue which are possessed by the human consciousness by dint or virtue of it being embodied.

However, human consciousness must not be in the misapprehension, miscalculation, confusion, illusion, wrong idea, or, false impression that, cosmic space or Brahmandic Aakash aka the consciousness of god or Brahman aka the mind of god or Brahman aka the creator of the physical-looking world,, <> (who is uniquely bodiless or, who uniquely, *does not exist* or *abide* in the physical-looking form, and therefore does not possess a physical-looking *speech-equipment,* consisting of lips and lungs plus larynx and tongue in the manner human consciousness possesses - by dint or virtue of its being embodied), <> does not communicate, converse, chat, speak, or talk, or, is not in communication, contact, or touch, or, does not have a conversation, or confab with the human consciousness in any way, shape, or form, or, in any possible manner.

On the contrary, the reality, is quite the opposite.

Cosmic space or Brahmandic Aakash aka the consciousness of god or Brahman aka the mind of god or Brahman aka the creator of the physical-looking world, <> (who is uniquely bodiless or, who uniquely *does not exist* or *abide* in physical-looking form, and therefore does not possess a physical-looking *speech-equipment,* consisting of lips and lungs plus larynx and tongue in the manner human consciousness possesses - by dint or virtue of it being embodied), <> is incessantly or relentlessly, communicating, conversing, chatting, speaking, or talking to, or, is in communication, contact, or in touch with, or, is having a conversation, or confab with human consciousness, both while human consciousness is in its wakeful state and while it is in its dream sleep state.

Let one explain what one means when one says that cosmic space or Brahmandic Aakash aka consciousness of god or Brahman aka mind of god or Brahman aka creator of the physical-looking world is incessantly or relentlessly communicating, conversing, chatting, speaking, or talking to, or, is in communication, contact, or in touch with, or, is having a conversation, or confab with human consciousness, both while the human consciousness is in its wakeful state and while it is in its dream sleep state.

Cosmic space or Brahmandic Aakash aka consciousness of god or Brahman aka mind of god or Brahman aka creator of the physical-looking world, is the source or spring of every idea or opinion occurring suddenly in the human consciousness or mind.

Cosmic space or Brahmandic Aakash aka consciousness of god or Brahman aka mind of god or Brahman aka creator of the physical-looking world, is the source or spring of each and every view, thought, concept, notion, impression, assumption, presumption, supposition, postulation, abstraction, line of thinking, belief, opinion, mental picture, hypothesis, theory, understanding, feeling, suspicion or hunch which erupts in the human consciousness or mind suddenly.

Similarly, cosmic space or Brahmandic Aakash aka consciousness of god or Brahman aka mind of god or Brahman aka creator of the physical-looking world, is the source or spring of each and every anticipation, expectation, prospect, contemplation, likelihood, possibility, fear, hope, aspiration, ambition, dream, intention, plan, design, purpose, aim which takes birth inside the human consciousness or mind suddenly.

Likewise, cosmic space or Brahmandic Aakash aka consciousness of god or Brahman aka mind of god or Brahman aka creator of the physical-looking world, is the source or spring of each and every action or process of thinking, reasoning, contemplation, musing, pondering, consideration, reflection, introspection, deliberation, study, rumination, cogitation, meditation, brooding, mulling over, concentration, debate, speculation or cerebration which takes place inside the human consciousness or mind.

In a similar way, cosmic space or Brahmandic Aakash

aka consciousness of god or Brahman aka mind of god or Brahman aka creator of the physical-looking world, is the source or spring of each and every emotion of concern for another's well-being, plus each and every feeling of compassion, sympathy, empathy, caring, concern, regard, consideration, kindness, tenderness, warmth, understanding, sensitivity, thoughtfulness, charity, benevolence, philanthropy, altruism or magnanimity that arise inside the human consciousness or mind.

By the same token, cosmic space or Brahmandic Aakash aka consciousness of god or Brahman aka mind of god or Brahman aka creator of the physical-looking world, is the source or spring of each and every dream which human consciousness or mind encounters and therefore, sees, observes, watches or perceives plus experiences during its dream-sleep state or during its rapid-eye-movement-sleep state, or, during its paradoxical-sleep state.

Cosmic space or Brahmandic Aakash aka consciousness of god or Brahman aka mind of god or Brahman aka creator of the physical-looking world, is the source or spring of all the above listed thoughts and more, which arise from time to time inside the human consciousness or mind.

It is also the source or spring of all kinds of dreams which take birth inside the human consciousness or mind during its dream-sleep state or during its rapid-eye-movement-sleep state, or, during its paradoxical-sleep state.

However, the brute fact is that the human consciousness or mind does not in the least bit or, does not in the slightest degree, sees eye to eye, or, agrees, accords, or, concurs with the statement that cosmic space or Brahmandic Aakash aka consciousness of god or Brahman aka mind of god or Brahman aka creator of the physical-looking world, is the source or spring of all thoughts that arise from time to time inside its being, that is to say, inside the human consciousness or mind.

Furthermore, human consciousness or mind also does not at all, or, does not in the slightest degree, agrees, accords, concurs or, sees eye to eye, with the statement that cosmic space or Brahmandic Aakash aka consciousness of god or Brahman aka mind of god or Brahman aka creator of the physical-looking world, is the source or spring of *dreams* that take birth inside it, that is to say, take birth inside the human consciousness or mind during latter's dream-sleep state or during latter's rapid-eye-movement-sleep state, or, during latter's paradoxical-sleep state.

Human consciousness or mind, instead, believes passionately or intensely that it itself is the source or spring of all thoughts that arise from time to time inside it and it also is the source or spring of all dreams that take birth inside it during its dream-sleep state or during its rapid-eye-movement-sleep state, or, during its paradoxical-sleep state.

The passionate or intense belief of human consciousness or mind, that it itself is the source or spring of all thoughts, that bubble up, from time to time, inside its being, plus, it also is the source or spring of all dreams that bubble up, inside its being, during its dream-sleep state or, during its rapid-eye-movement-sleep state, or, during its paradoxical-sleep state, <> (and therefore, all this bubbling up, inside its being, of thoughts on one hand and dreams on the other, has nothing to do with cosmic space or Brahmandic Aakash aka consciousness of god or Brahman aka mind of god or Brahman aka creator of the physical-looking world), <> is due to the *cosmic delusion* or *brahmandic vibhranti* from which human consciousness suffers, or, is afflicted by, or, is troubled with.

This *cosmic delusion* or *brahmandic vibhranti* from which human consciousness suffers or, is afflicted by, or, is troubled with, has been bestowed or conferred on the human consciousness or mind by none other than cosmic space or Brahmandic Aakash aka consciousness of god or Brahman aka mind of god or Brahman aka creator of the physical-looking world.

The only being who can free human consciousness or mind from the iron grip, clutch, hold, or, embrace of this *cosmic delusion* or *brahmandic vibhranti* is cosmic space or Brahmandic Aakash itself and no one else.

That is to say, the only being who can free human consciousness or mind from the iron grip, clutch, hold, or, embrace of the *cosmic delusion* or *brahmandic vibhranti* in question, is consciousness of god or Brahman or, mind of god or Brahman aka creator of the physical-looking world, itself and no one else.

This is so because cosmic space or Brahmandic Aakash aka consciousness of god or Brahman aka mind of god or Brahman aka creator of the physical-looking world, is the bestower or conferrer of the *cosmic delusion* or *brahmandic vibhranti* in question, on the human consciousness or mind and therefore, only it has the power to free human consciousness or mind from the iron grip, clutch, hold, or, embrace of the *cosmic delusion* or *brahmandic vibhranti* in question and no one else.

The reason why cosmic space or Brahmandic Aakash aka consciousness of god or Brahman aka mind of god or Brahman aka creator of the physical-looking world, has bestowed or conferred on the human consciousness or mind, the *cosmic delusion* or *brahmandic vibhranti* in question, is the following :-

Without the formidable veil of the *cosmic delusion* or *brahmandic vibhranti* in question, on the analytical or discriminatory faculty of human consciousness, the latter will never enjoy the magic show called the physical world i.e., the physical-looking world and will not become submerged into it with every ounce of its body and soul as it is doing at present or, as is its wont or habit, at present.

This formidable veil called *cosmic delusion* or *brahmandic vibhranti,* shrouding or befogging the analytical or discriminatory faculty of human consciousness, is vital for the smooth functioning of the magic show called the physical world i.e., the physical-looking world. In this context, human consciousness must always bear in mind the following truth :-

The sole purpose of cosmic space or Brahmandic Aakash aka God or Brahman, behind the creation of its imagery, dreamry, daydream or oneiric which is unenlightenedly or nesciently called the real-physical world by the human consciousness (which, in fact, is not in the least either real or physical, but is one hundred percent daydreamal or oneirickle in nature or quintessence) is to derive personal fun, joy, amusement, entertainment or regalement out of seeing, observing, watching, beholding or perceiving it.

Human consciousness must also remember that cosmic

space or Brahmandic Aakash aka God or Brahman is the magician whose magic show, the physical world is i.e., the physical-looking world is.

CONSCIOUSNESSBAL COSMOLOGY VS PHYSICAL COSMOLOGY.

INTRODUCTION.

The expression "cosmology" can be defined as the study of origin, fundamental nature, structure, natural order, development or evolution, and eventual fate of the universe. That is to say, the term "cosmology" encompasses the entire life-span of the universe from birth to death, with puzzles or enigmas at every stage of its evolution or development.

Cosmology, as defined above, can be studied from two very different perspectives or angles, namely, the consciousnessbal, awarenessbal, or, sentiencel perspective or angle, on one hand, and the *physical, material, substantial, objective, or, concrete* perspective or angle, on the other.

Looking at the cosmos from these two very different perspectives is mandatory or compulsory for the human consciousness if human consciousness is really very keen or serious to know the absolutely true origin, fundamental nature, structure, natural order, development or evolution, and eventual fate of the universe. This is the irrefutable, incontestable, or undeniable case, with regards to this issue because cosmos is composed, not merely of one, but two ingredients, namely, consciousness, on one hand, and, physical matter, on the hand.

Hence, "cosmology" can be classified as being of two basic kinds, namely, consciousnessbal cosmology, on one hand, and physical cosmology, on the other.

Present-day, physical cosmology is based largely on the *material theory* of Big Bang and the associated *material thing* or *entity* called or named singularity or cosmic egg, (inside which this event of Big Bang is theorized to have taken place by the votaries or believers of this theory, some 13.7 billion light years ago, so as to give birth to current material, substantial, physical, objective, or concrete cosmos), as well as, yet to be found, such *material things* or *entities* as dark matter and dark energy.

The attempt of the *material theory* of Big Bang is to bring together, the two very well-known branches of *material science*, namely, that of the particle physics on one hand, and that of the observational astronomy on the other.

The physical cosmology does not explain how the *birth of consciousness* and the *birth of cosmic space* took place in the current, material, substantial, physical, objective, or concrete cosmos or universe. It merely conjectures that the *consciousness* and *cosmic space* are both purely two very insignificant or unimportant, fruits, spin-offs, by-products, incidental-products, or, unintended but inevitable, secondary products, made during the manufacture or synthesis of physical cosmos out of the supreme truth of the current physical or concrete cosmos, namely, the physical matter of the current cosmos. That's all.

In contrast, the viewpoint of the consciousnessbal cosmology, (which consists of Adwait-Vedantic cosmology), about the origin, fundamental nature, structure, natural order, development or evolution, and eventual fate of the universe is absolutely contrarian, antipodean, or diametrically opposite to the viewpoint of the physical cosmology.

The consciousnessbal cosmology of Adwait-Vedanta deals with the issue of origin, fundamental nature, structure, natural order, development or evolution, and eventual fate of the universe from the consciousnessbal aspect and not from the physical aspect of the universe. But it does clearly explain how physical matter has been created in the universe out of consciousness and how consciousness and cosmic space themselves, have both come into being in the universe.

The consciousnessbal cosmology of Adwait-Vedanta deals with the issue of *birth of the physical matter* in the universe as well as the issue of the presence of consciousness plus the issue of the presence of cosmic space in the universe, head-on, and without any dilly-dally, and obfuscation, from the pure perspective of consciousness and nothing else.

The consciousnessbal cosmology of Adwait-Vedanta uses human consciousness's own daily experience of dream sleep state at night, on one hand, and its self-willed, experience of daydreaming or oneiricking, while it is in its wakeful state, on the other, as two very *direct* or *personal* experiential tools to expound or, to present and explain itself in detail or, to spell out, describe, discuss, delineate, explicate or elucidate itself in detail.

The consciousnessbal cosmology of Adwait-Vedanta uses these two very *direct* or *personal* experiential tools, viz., human consciousness's own daily experience of dream sleep state at night, on one hand, and its self-willed, experience of daydreaming or oneiricking, while it is in its wakeful state, on the other; both of which are in the range, reach, scope, capacity, or compass of each and every human consciousnesses, to answer questions regarding the origin, fundamental nature, structure, natural order, development or evolution, and eventual fate of the universe from the consciousnessbal aspect and not from the physical aspect of the universe.

PHYSICAL COSMOLOGY.

The material science of today passionately believes that the *physical matter* of the contemporary cosmos is the supreme truth of the current cosmos and the remaining two items, which are extant, around or present in the current cosmos, namely, *cosmic space,* on one hand, and *consciousness,* on the other, for example, human *consciousness,* are less important truths.

That is to say, as per material science of today, the *physical matter* of the contemporary cosmos is the supreme truth of the current cosmos and the remaining two items, which are extant, around or present in the current cosmos, namely, *cosmic space* on one hand, and *consciousness* on the other, for example, human *consciousness,* are both subordinate, subservient, second-fiddle, second-class, secondary, subsidiary, supplementary, supplemental, peripheral, lesser, minor, auxiliary, ancillary, lower-level, or, lower-grade truths because they both have emanated or originated from, or, they both have been disgorged into the current cosmos or have been given off into the current cosmos by the supreme truth of the current cosmos namely the *physical matter* of the current cosmos.

What has been said above can be put in another way.

Cosmic space, on one hand, and human consciousness, on the other, of the contemporary cosmos, are both regarded by the material science of today as being a pair of such insignificant entities which are too small or unimportant to be worth consideration in the scheme of things called the physical matter, on one hand, and the physical cosmos, on the other.

In other words, cosmic space and consciousness, for example, human consciousness of the contemporary cosmos, are both regarded by the material science of today as being a pair of insignificant or unimportant, fruits, spin-offs, by-products, incidental-products, or, unintended but inevitable, secondary products, made during the manufacture or synthesis of physical cosmos out of the supreme truth of the current cosmos, namely, the physical matter of the current cosmos.

The term *"physical cosmology"* encompasses the above sentiment or viewpoint of the contemporary science.

The term *"physical cosmology"* conveys or expresses the sentiment or viewpoint of the present-day material science that every truth or item or, every being or thing, which is extant, around or present in the current cosmos, has emanated from, or has issued forth from or, has come into being in the current cosmos from the physical matter of the current cosmos and nothing else.

The word "everything" includes cosmic space of the cosmos on one hand, and consciousness of the cosmos, on the other, for example, human consciousness of the cosmos.

Therefore, as per material science of today, the physical matter of the current cosmos is the *numero uno, number one, first and last* or the *leader* of the current cosmos and not cosmic space or human consciousness.

CONSCIOUSNESSBAL COSMOLOGY.

Standing in opposition to the above sentiment or viewpoint of the contemporary material science or, opposing or confronting the above sentiment or viewpoint of the contemporary material science is the Adwait-Vedantic viewpoint or standpoint which holds the view that in the contemporary cosmos, it is *cosmic space* alone which is the supreme reality because this cosmic space is an incredible or extraordinary plus ubiquitous and infinite field of consciousness of god aka the source or spring of *consciousness* in the current cosmos including human consciousness, aka the creator of the physical matter in the current cosmos and therefore, the creator of the present-day physical cosmos itself.

The Adwait-Vedantic viewpoint is that cosmic space, which is seen by the human consciousness via the media of the eyes of its physical body during its wakeful state <> (and not during its dream sleep state) <> is not an unremarkable, humdrum, mundane, colorless, dull, boring, lacklustre, uninteresting or uninspiring thing. Nor is it an existence, being, truth or reality which lacks or is devoid of unusual or special mystery, enigma, conundrum, riddle, puzzle or secret, buried deep inside it.

On the contrary, cosmic space, which is seen by the

human consciousness with the help of the eyes of its physical body during its wakeful state, is the consciousness of god or the mind of god. This consciousness of god or the mind of god aka cosmic space is the creator of the physical-looking matter and therefore, the creator of the physical-looking world, cosmos or universe which human consciousness encounters, perceives, experiences and takes part in, during its wakeful state. Inside this consciousness or mind of god aka cosmic space, the physical-looking world, cosmos or universe; consisting of countless physical-looking moons, planets, stars, black holes and galaxies; is floating, wafting, or levitating, plus whirling, twirling or spiraling nonstop and has been doing so from the beginning of the current cosmic time, which was some 13.7 billion light years ago, and will continue to do so till the end of the current cosmic time.

The term *"consciousnessbal cosmology"* encompasses the above sentiment or viewpoint of Adwait-Vedanta.

Therefore, the term *"consciousnessbal cosmology"* highlights, brings to the fore, draws special attention to or underscores the viewpoint of Adwait-Vedanta that it is the consciousness in the current cosmos, namely, cosmic space which is the *numero uno, number one, first and last* or the leader of the current cosmos and not physical matter.

That is to say, the term *"consciousnessbal cosmology"* highlights, brings to the fore, draws special attention to or underscores the viewpoint of *Adwait-Vedanta* that it is the cosmic space which is the supreme reality in the current cosmos because it is an incredible or extraordinary plus ubiquitous and infinite field of consciousness of god aka the source or spring of consciousness in the current cosmos including human consciousness aka the creator of the *physical matter* in the current cosmos and therefore, the creator of the present-day *physical cosmos* itself.

Hence the term *"consciousnessbal cosmology"* conveys or expresses the sentiment or the viewpoint of Adwait-Vedanta that everything in the current cosmos has emanated from or, has issued forth from or, has come into being from the incredible or extraordinary plus ubiquitous and infinite field of consciousness called cosmic space who is none other than god aka the source or spring of all consciousnesses in the current cosmos including all human consciousnesses aka the creator of the physical matter in the current cosmos and therefore, the creator of the present-day physical cosmos itself.

DREAMS PERCEIVED BY HUMAN CONSCIOUSNESS DURING DREAM SLEEP STATE EXIST INSIDE HUMAN CONSCIOUSNESS BUT ARE CREATED BY COSMIC SPACE.

Human beings think it likely that, the *dreams* of their *dream sleep state* are created, generated or produced by their own consciousness, during their *dream sleep state.*

But the above, is not at all the case with regards to the *dreams* of the *dream sleep state* of human beings. The latter's above assumption, belief, conjecture, guess-work, hunch, or hypothesis with regards to the *dreams* of their *dream sleep state* is one hundred percent wrong. The fact, reality or truth with regards to the *creation*, *generation* or *production* i.e., *genesis,* or *origin,* of *dreams* during human being's *dream sleep state* is quite different.

This fact, reality or truth with regards to the *creation*, *generation* or *production* i.e., *genesis,* or *origin,* of *dreams* during human being's *dream sleep state* will be described a bit later but before that, let one describe what actually, or truly happens during *dreams* of human being's *dream sleep state*.

The *dream* of each and every human being's *dream sleep state* is met, faced, encountered, plus experienced by two twins, clones, incarnations, or avatars of this human being's consciousness.

One twin, clone, incarnation, or avatar of human being's consciousness during latter's dream sleep state, finds itself being *subjectively* bodiless while the other twin, clone, incarnation, or avatar of human being's consciousness during latter's dream sleep state, finds itself being *objectively* embodied.

The twin, clone, incarnation, or avatar of human being's consciousness during latter's dream sleep state, which finds itself being *objectively* embodied, is made to wear or put on, or, is made to cloth, enrobe, or drape itself with a consciousnessbal-substance-made or, consciousnessbal-matter-made *body* which possesses five consciousnessbal-substance-made or, five consciousnessbal-matter-made *sense organs* for its exclusive use during the dream sleep state of human consciousness.

The *objectively* embodied *twin, clone, incarnation* or *avatar* of human being's consciousness during latter's dream sleep state, (which is made to wear or put on, or, is made to cloth, enrobe, or drape itself with, a consciousnessbal-substance-made or consciousnessbal-matter-made *body*), observes, sees, visualises or watches, plus, feels or touches, and hears, smells, and tastes, that is to say, to the fullest extent perceives and experiences as well as participates in, or, takes part in all the *dreamal* or *phantasmal activities* of the *dreamal* or *phantasmal* world, cosmos or universe of human being's *dream sleep state*, with the assistance of its consciousnessbal-substance-made or consciousnessbal-matter-made *body* and its body's five consciousnessbal-substance-made or consciousnessbal-matter-made *sense organs* during human being's *dream sleep state.*

In sharp, stark, utter, complete, or, absolute contrast to the *objectively* embodied, *twin, clone, incarnation* or *avatar* of human being's consciousness during latter's dream sleep state, (which is made to wear or put on, or, is made to cloth, enrobe, or drape itself with, a consciousnessbal-substance-made or consciousnessbal-matter-made *body*), the *subjectively* bodiless, *twin, clone, incarnation, or avatar* of human being's consciousness during latter's dream sleep state, possesses no body from the standpoint of its own *subjective* awareness, apprehension, perception, appreciation, cognisance, feeling, knowledge, realisation, understanding, grasp, or frame of reference, or, is without a body from the standpoint of its own *subjective* awareness, apprehension, perception, appreciation, cognisance, feeling, knowledge, realisation, understanding, grasp, or frame of reference, even though it factually continues to wear or don the *physical* or *objective* body of its wakeful state throughout its dream sleep state.

The *subjectively* bodiless, *twin, clone, incarnation, or avatar* of human being's consciousness during latter's dream sleep state, merely, purely or simply, observes, sees, visualises or watches, plus to that limited extent, i.e., to the limited extent of merely, purely or simply observing, seeing, visualising or watching, it also experiences each and every *dreamal* or *phantasmal activity* plus each and every *dreamal* or *phantasmal* being, which materialises, concretises or, *objectifies* inside the *dreamal* or *phantasmal* world, cosmos or universe of human being's *dream sleep state.*

The expression "each and every *dreamal* or *phantasmal activity* plus each and every *dreamal* or *phantasmal* being which materialises, concretises or, *objectifies* in the *dreamal* or *phantasmal* world, cosmos or universe of human being's *dream sleep state"* includes the *objectively* embodied *twin, clone, incarnation* or *avatar* of human being's consciousness which from time to time materialises, concretises or, *objectifies* in the *dreamal* or *phantasmal* world, cosmos or, universe during latter's dream sleep state, and all the *dreamal* or *phantasmal activities* which this *objectively* embodied, *twin, clone, incarnation* or *avatar* of human being's consciousness during latter's dream sleep state, engages in, becomes involved in, take parts in or participate in, or, carries out, discharges, executes, performs, or undertakes inside this *dreamal* or *phantasmal* world, cosmos or universe of human being's *dream sleep state.*

The *subjectively* bodiless, *twin, clone, incarnation, or avatar* of human being's consciousness during latter's dream sleep state, is not in its usual or customary, dimensionless form or version.

Instead, it is, in its unusual or uncommon 3-D or three-dimensional form or version or, in its unusual or uncommon expanded, distended, dilated, or inflated form or version because the 3-D or three-dimensional *dream,* which it observes, sees, visualises, or watches during its dream sleep state, is situated, sited, or stationed inside it and not outside it. And, in order to spatially accommodate inside itself this 3-D or three-dimensional *dream,* which it observes, sees, visualises or watches during its dream sleep state, it itself has to become, of necessity, perforce, inevitably, unavoidably, or by force of circumstances, 3-D or three-dimensional in configuration or silhouette, by unawarely or unconsciously, expanding, distending, dilating or inflating itself.

What has been said above, can be put in another way.

The *subjectively* bodiless, *twin, clone, incarnation, or avatar* of human being's consciousness during its dream sleep state, unawarely or unconsciously, exists, occurs, obtains, or, is in existence in its unusual or uncommon, 3-D or three-dimensional form or version or, in its unusual or uncommon, expanded, distended, dilated, or inflated form or version, throughout the total time, spell or span of the existence or presence of its *dream sleep state's, dreamal* or *phantasmal* world, cosmos or universe inside itself, which is quite contrary to its usual or customary way, wont, habit, practice, custom, convention, or routine during its wakeful state.

The usual or customary way, wont, habit, practice, routine or custom of human consciousness during its wakeful state is to, unawarely or unconsciously, exist, occur, obtain, or, be in existence majority of the time in its boring, colourless, dull, humdrum, mundane, average, ordinary, or run-of-the-mill, dimensionless form or version, or, in its standard, regular or conventional, unexpanded, undistended, undilated, or uninflated form or version except when some times or on on some occasions, it daydreams or oneirics.

That is to say, the usual or customary way, wont, habit, practice, custom, convention, routine, or, rule of human being's consciousness during its wakeful state is to exist or abide, unawarely or unconsciously, most of the time in its standard, regular, routine or common, dimensionless form or version or, in its standard, regular, routine or common, unexpanded, undistended, undilated, or uninflated form or version and not in its unusual or uncommon, 3-D or three-dimensional form or version i.e., and not in its unusual or uncommon, expanded, distended, dilated, or inflated form or version.

Human consciousness during its wakeful state, only occasionally, uncommonly, or once in a while exists or abides in its 3-D or three-dimensional form or version or, in its expanded, distended, dilated, or inflated form or version.

Human consciousness during its wakeful state exists or

abides in its 3-D or three-dimensional form or version or, in its expanded, distended, dilated, or inflated form or version only when it sometimes, consciousnessbally, awarenessbally or mentally conceptualises, visualises, fantasises, conceives, perceives, or imagines.

That is to say, human consciousness during its wakeful state exists or abides in its 3-D or three-dimensional form or version or, in its expanded, distended, dilated, or inflated form or version only when it sometimes or, on rare occasion, daydreams or oneirics.

To reiterate.

Human consciousness during its wakeful state, abides or exists in its 3-D or three-dimensional form or version or, in its expanded, distended, dilated, or inflated form or version, only very rarely. For example, it abides or exists in its 3-D or three-dimensional form or version or, in its expanded, distended, dilated, or inflated form or version during its wakeful state, only when, sometimes, or, on rare occasion, it consciousnessbally, awarenessbally or mentally, conceptualises, visualises, fantasises, conceives, perceives, or imagines, or, daydreams or oneirics.

Otherwise, unawaredly or unconsciously, it stays put, holds fast, or clings to its usual or customary, dimensionless form or version or, to its usual or customary unexpanded, undistended, undilated, or uninflated form or version.

When sometimes, or on rare occasion, human consciousness, in its wakeful state, begins, starts, undertakes, engages in, or, enters into, its activity of daydreaming or oneiricking or, into its activity of consciousnessbal, or awarenessbal conceptualisation, visualisation or fantasisation, it becomes or, it transforms, transmutes, or, metamorphoses itself, unawaredly or unconsciously, into its unusual or uncommon 3-D or three-dimensional form or version or, into its unusual or uncommon, expanded, distended, dilated, or inflated form or version. Otherwise, at all other times, it unawarely or unconsciously, abides or exists in its usual or customary, dimensionless form or version or, in its usual or customary, unexpanded, undistended, undilated, or uninflated form or version.

The unusual or uncommon, 3-D or three-dimensional form or version or, the unusual or uncommon, expanded, distended, dilated, or inflated form or version of the *subjectively* bodiless, *twin, clone, incarnation, or avatar* of human being's consciousness during its dream sleep state, is alike, akin, similar, or, identical, in attributes, features, qualities, traits, character, nature, role, or, part, to the unusual or uncommon, 3-D or three-dimensional form or version or, the unusual or uncommon, expanded, distended, dilated, or inflated form or version of God's or Brahman's consciousness, viz., of the consciousness of the Daydreamer, Oneiricker, Author, Maker, or Creator, or, the consciousness of the Progenitor, Primogenitor, or Begetter, of the *physical, material,* or, *concrete* cosmos, world, or, universe. The unusual or uncommon, 3-D or three-dimensional form or version or, the unusual or uncommon, expanded, distended, dilated, or inflated form or version of God's or Brahman's consciousness, viz., of the consciousness of the Daydreamer, Oneiricker, Author, Maker, or Creator, or, of the consciousness of the Progenitor, Primogenitor, or Begetter, of the *physical, material,* or, *concrete* cosmos, world, or, universe, which is observed, seen, visualised, or watched by the human being's consciousness during its wakeful state with the help of the eyes of its physical, material, or, concrete body of its wakeful state, is called, named, addressed, designated, dubbed, titled, or, tagged, by the latter, namely, by the human being's consciousness during its *wakeful state,* as Cosmic Space or Brahmandic Aakash.

Cosmic Space or Brahmandic Aakash, that is to say, the unusual or uncommon, 3-D or three-dimensional form or version or, the unusual or uncommon, expanded, distended, dilated, or inflated form or version of God's or Brahman's consciousness, viz., of the consciousness of the Daydreamer, Oneiricker, Author, Maker, or Creator, or, of the consciousness of the Progenitor, Primogenitor, or Begetter, of the *physical, material,* or, *concrete cosmos, world, or, universe,* can only be observed, seen, visualised or watched by human being's consciousness during its wakeful state, with the help of the eyes of the physical, material, or, concrete body of its wakeful state.

However, Cosmic Space or Brahmandic Aakash, that is to say, the unusual or uncommon, 3-D or three-dimensional form or version or, the unusual or uncommon, expanded, distended, dilated, or inflated form or version of God's or Brahman's consciousness, viz., of the consciousness of the Daydreamer, Oneiricker, Author, Maker, or Creator, or, of the consciousness of the Progenitor, Primogenitor, or Begetter of the *physical, material,* or, *concrete cosmos, world, or, universe,* cannot be felt, palpated, or, touched by the hands, feet, or, by any other part of the physical body of human being's consciousness during its wakeful state.

That is to say, Cosmic Space or Brahmandic Aakash, i.e., the unusual or uncommon, 3-D or three-dimensional form or version or, the unusual or uncommon, expanded, distended, dilated, or inflated form or version of God's or Brahman's consciousness, viz., of the consciousness of the Daydreamer, Oneiricker, Author, Maker, or Creator, or, the consciousness of the Progenitor, Primogenitor, or Begetter of the *physical, material,* or, *concrete cosmos, world, or, universe,* cannot be felt, or, touched by the skin of the physical body of human being's conscious during its wakeful state.

Furthermore, Cosmic Space or Brahmandic Aakash, that is to say, the unusual or uncommon, 3-D or three-dimensional form or version or, the unusual or uncommon, expanded, distended, dilated, or inflated form or version of God's or Brahman's consciousness, viz., of the consciousness of the Daydreamer, Oneiricker, Author, Maker, or Creator, or, of the consciousness of the Progenitor, Primogenitor, or Begetter of the *physical, material,* or, *concrete cosmos, world, or, universe,* cannot be tasted, heard, or smelled by the tongue, ears, and nose of the physical, material, or, concrete body of human being's consciousness during its wakeful state.

Thus, the existence or presence of Cosmic Space or Brahmandic Aakash, that is to say, the existence or presence of the unusual or uncommon, 3-D or three-dimensional form or version or, the unusual or uncommon, expanded, distended, dilated, or inflated form or version of God's or Brahman's consciousness, viz., of the consciousness of the Daydreamer, Oneiricker, Author, Maker, or Creator, or, the consciousness of the Progenitor, Primogenitor, or Begetter of the *physical, material,* or, *concrete cosmos, world, or, universe* can be recognised, appreciated, apprehended, comprehended, grasped, discerned, detected, figured out, made out, verified, or, spotted, with the help of only one *sense organ,* namely, the eyes of the physical body of human being's consciousness during latter's wakeful state and not by the remaining four *sense organs,* namely, the skin, tongue, nose, and ears.

But the fact remains, or, better still, the unsentimental-fact, unvarnished-fact, down-to-earth-fact, or, hard-boiled-fact, is that, the existence or presence of the Daydreamer, Oneiricker, Author, Maker, or Creator, or, of the Progenitor, Primogenitor, or Begetter of the *physical, material,* or, *concrete cosmos, world, or, universe*, called Cosmic Space or Brahmandic Aakash, can be recognised, appreciated, apprehended, comprehended, grasped, discerned, detected, made out, verified, or, spotted, by human being's consciousness during its wakeful state, albeit with the help of only one, and one only *sense organ,* namely, the eyes of its physical body during its wakeful state, and not with the help of the remaining four *sense organs* of its physical body.

In other words, the Daydreamer, Oneiricker, Author, Maker, or Creator, or, the Progenitor, Primogenitor, or Begetter, of the *physical, material,* or, *concrete cosmos, world, or, universe* is not absolutely or, totally beyond the ken, knowledge, awareness, appreciation, apprehension, comprehension, grasp, recognition, realisation, or, understanding of all the *sense organs of the* physical body of human being's consciousness during its wakeful state as is widely believed, thought, imagined, maintained, assumed, presumed, conjectured, surmised, postulated, theorised, concluded, deduced, guessed, or, reckoned by most human being's consciousnesses.

Now let one pick up the original thread once again from where one left it earlier.

Inside the ubiquitous and infinite field of consciousness called Cosmic Space or Brahmandic Aakash, which is the unusual or uncommon, 3-D or three-dimensional form or version or, the unusual or uncommon, expanded, distended, dilated, or inflated form or version of God's or Brahman's consciousness, viz., of the consciousness of the Daydreamer, or, Oneiricker, or, of the consciousness of the Author, Maker, or Creator, or, of the consciousness of the Progenitor, Primogenitor, or Begetter, of the *physical, material,* or, *concrete cosmos, world, or, universe,* the latter, namely, the physical, material, or, concrete cosmos, world, or, universe, which comprises or consists of countless physical, material, or, concrete moons, planets, stars, black holes, galaxies and the like, is floating, wafting, or levitating, plus whirling, twirling or spiralling nonstop, as a mere daydream or oneiric, or, as a mere dreamry, imagery, or fantasy of its

Daydreamer, or, Oneiricker, or, of its Author, Maker, or Creator, or, of its Progenitor, Primogenitor, or Begetter, which is none other than the same, very same, or selfsame, ubiquitous and infinite field of consciousness called Cosmic Space or Brahmandic Aakash, which, in turn, is nothing but the unusual or uncommon, 3-D or three-dimensional form or version or, the unusual or uncommon, expanded, distended, dilated, or inflated form or version of God's or Brahman's consciousness, viz., of the consciousness of the Daydreamer, or, Oneiricker, or, of the consciousness of the Author, Maker, or Creator, or, of the consciousness of the Progenitor, Primogenitor, or Begetter, of the *physical, material,* or, *concrete cosmos, world, or, universe.*

To paraphrase.

The physical, material, or, concrete cosmos, world, or, universe, which comprises or consists of countless physical, material, or, concrete moons, planets, stars, black holes, galaxies and the like, has been floating, wafting, or levitating, plus whirling, twirling or spiralling nonstop, as a mere daydream or oneiric, or, as a mere dreamry, imagery, or fantasy and nothing else, inside the ubiquitous and infinite field of consciousness called Cosmic Space or Brahmandic Aakash.

The ubiquitous and infinite field of consciousness called

Cosmic Space or Brahmandic Aakash is the unusual or uncommon, 3-D or three-dimensional form or version, or, the unusual or uncommon, expanded, distended, dilated, or inflated form or version of the consciousness of the Daydreamer, or, Oneiricker, or, of the consciousness of the Author, Maker, or Creator, or, of the consciousness of the Progenitor, Primogenitor, or Begetter, or, of the consciousness of the Source or Spring, or, of the consciousness of God or Brahman, whose daydream or oneiric, or, whose dreamry, imagery, or fantasy, the physical, material, or, concrete cosmos, world, or, universe is.

Hence, the physical, material, or, concrete cosmos, world, or, universe, which comprises or consists of countless physical, material, or, concrete moons, planets, stars, black holes, galaxies and the like, is floating, wafting, or levitating, plus whirling, twirling or spiralling nonstop as a mere daydream or oneiric and nothing else, or, as a mere dreamry, imagery, or fantasy and nothing else, inside the ubiquitous and infinite field of consciousness called Cosmic Space or Brahmandic Aakash, which is the unusual or uncommon, 3-D or three-dimensional form or version, or, the unusual or uncommon, expanded, distended, dilated, or inflated form or version of the consciousness of the Daydreamer, or, Oneiricker, or, of the consciousness of the Author, Maker, or Creator, or, of the consciousness of the

Progenitor, Primogenitor, or Begetter, or, of the consciousness of the Source or Spring, or, of the consciousness of God, Brahman, whose daydream or, oneiric, the physical, material, or, concrete cosmos, world, or, universe is.

In other words, physical, material, or, concrete cosmos, world, or, universe is an incredible daydream or oneiric, or, an incredible dreamry, imagery, or, fantasy of its Daydreamer, or, Oneiricker, or, of its Author, Maker, Creator, Progenitor, Primogenitor, Begetter, Source, Spring, God, or, Brahman.

And this astounding, astonishing, amazing, or, awe-inspiring, Daydreamer, or, Oneiricker, or, Author, Maker, Creator, Progenitor, Primogenitor, Begetter, Source, Spring, God, or, Brahman of the physical, material, or, concrete cosmos, world, or, universe is the ubiquitous and infinite field of consciousness called Cosmic Space or Brahmandic Aakash, which in turn, is the unusual or uncommon, 3-D or three-dimensional form or version, or, the unusual or uncommon, expanded, distended, dilated, or inflated form or version of the consciousness of the Daydreamer, or, Oneiricker, or, of the consciousness of the Author, Maker, Creator, Progenitor, Primogenitor, Begetter, Source, Spring, God, or, Brahman, whose daydream, or, oneiric, or, whose dreamry, imagery, or fantasy, the current physical, material, or, concrete cosmos, world, or, universe is.

What has been said above can be put in another way.

The physical, material, or, concrete *cosmos, world, or, universe* which human being's consciousness in its physically, materially, or, concretely *embodied* form or version, during its wakeful state, perceives and experiences, willy-nilly, perforce, without a choice, whether desired or not, or, every which way, each and every day, plus, in which it participates or, takes part, willy-nilly, perforce, without a choice, whether desired or not, or, every which way, each and every day, during its wakeful, is a mere daydream or oneiric, or, is a mere dreamry, imagery, or fantasy, of the consciousness of its daydreamer, or, oneiricker, or, the consciousness of its Author, Maker, or Creator, or, the consciousness of its Progenitor, Primogenitor, or Begetter, or, the consciousness of its Source or Spring, or, the consciousness of its God or Brahman, which is none other than the astounding, astonishing, amazing, or, awe-inspiring, *being, truth, reality,* or, *actuality,* called the ubiquitous and infinite field of consciousness aka Cosmic Space or Brahmandic Aakash, which, in turn, is nothing but the unusual or uncommon, 3-D or three-dimensional form or version or, the unusual or uncommon, expanded, distended, dilated, or inflated form or version of God's or Brahman's consciousness, viz., of the consciousness of the daydreamer, or, onciricker, or, of the consciousness of the Author, Maker, or Creator, or, of the consciousness of the Progenitor, Primogenitor, or Begetter, or, of the consciousness of the Source or, Spring of the *physical, material,* or, *concrete cosmos, world, or, universe.*

It is very essential for one to understand clearly and decidedly the true nature of Cosmic Space or Brahmandic Aakash, on one hand, and the true nature of the physical, material, or, concrete cosmos, world, or, universe, on the other.

Therefore, it will be beneficial for one to describe once again, the true nature of Cosmic Space or Brahmandic Aakash as well as the true nature of the physical, material, or, concrete cosmos, world, or, universe.

It will also be beneficial for one to point out once again, the true nature of the relationship that obtains or exists between the Cosmic Space or Brahmandic Aakash, on one hand, and the physical, material, or, concrete, cosmos, world, or, universe, on the other.

The physical, material, or, concrete, cosmos, world, or, universe, which human being's consciousness, in its physically, materially, or, concretely *embodied* form or version, during its wakeful state, perceives and experiences, willy-nilly, perforce, without a choice, whether desired or not, or, every which way, each and every day, plus, in which it participates, or, takes part, during its wakeful state, willy-nilly, perforce, without a choice, whether desired or not, or, every which way, each and every day, is nothing but a mere daydream or oneiric, or, a mere dreamry, imagery, or fantasy of the consciousness of its daydreamer, or, oneiricker, or, the consciousness of its Author, Maker, or Creator, or, the consciousness of its Progenitor, Primogenitor, or Begetter, who is none other than the astounding, astonishing, amazing, or, awe-inspiring, *being, truth, actuality,* or, *reality,* called the ubiquitous and infinite field of consciousness aka Cosmic Space or Brahmandic Aakash, which is observed, seen, visualised, or, watched by the human being's consciousness in its physically, materially, or, concretely *embodied* form or version during its wakeful state, willy-nilly, perforce, without a choice, whether desired or not, or, every which way, each and every day, and inside which all the physical, material, or, concrete moons, planets, stars, black holes, galaxies and the like, are floating, wafting, or levitating, plus whirling, twirling or spiralling nonstop, and has been doing so from the beginning of *cosmic time* and will continue to do so till the end of *cosmic time,* willy-nilly, perforce, without a choice, whether desired or not, or, every which way, each and every day.

Just to remind one, once again that the physical, material,

or, concrete cosmos, world, or, universe is the cosmos, world, or, universe which human being's consciousness, in its physically, materially, or, concretely, embodied *form* or *version* during its wakeful state, willy-nilly, perforce, without a choice, whether desired or not, or, every which way, each and every day, observes, sees, visualises, watches, feels, palpates, touches, tastes, hears, and smells, that is to say, perceives & experiences, *fully, completely, all out,* or, *in all respects,* i.e., *to the fullest extent,* plus, in which it willy-nilly, perforce, without a choice, whether desired or not, or, every which way, each and every day, participates or takes part, *fully, completely, all out,* or *in all respects,* i.e., *to the fullest extent,* through the *agency, medium,* or, *instrumentality* of its physical, material, or, concrete body, on one hand, and through the *agency, medium,* or, *instrumentality* of the *five* sense organs of its physical, material, or, concrete body, on the other.

What has been said above can be put in another way.

The ubiquitous and infinite field of consciousness called Cosmic Space or Brahmandic Aakash is the unusual or uncommon, 3-D or three-dimensional form or version or, the unusual or uncommon, expanded, distended, dilated, or inflated form or version of the usually or customarily dimensionless form or version or, the usually or customarily unexpanded, undistended, undilated, or uninflated form or version of the *consciousness* of God or Brahman, or, the *consciousness* of the Author, Maker, or, Creator, or, the *consciousness* of the Progenitor, Primogenitor, or Begetter, or, the *consciousness* of the Source or, Spring, of the physical, material, or, concrete cosmos, world, or, universe.

And, inside this ubiquitous and infinite field of consciousness called Cosmic Space or Brahmandic Aakash, which is the unusual or uncommon, 3-D or three-dimensional form or version or, the unusual or uncommon, expanded, distended, dilated, or inflated form or version of the usually or customarily dimensionless form or version or, the usually or customarily unexpanded, undistended, undilated, or uninflated form or version of the *consciousness* of the daydreamer, or, oneiricker, or, the consciousness of the Author, Maker, or, Creator, or, the *consciousness* of the Progenitor, Primogenitor, or Begetter, or, the *consciousness* of the Source or, Spring, or, the consciousness of God or Brahman, of the physical, material, or, concrete, cosmos, world, or, universe, the latter, namely, the physical, material, or, concrete, cosmos, world, or universe, consisting of countless moons, planets, stars, black holes, galaxies, and the like, which human being's consciousness, in its physically, materially, or, concretely, embodied *form* or *version,* during its wakeful state, perceives, and experiences, willy-nilly, perforce, without a choice, whether desired or not, or, every which way, each and every day, *fully, completely, all out,* or, *in all respects,* i.e., *to the fullest extent,* and in which, human being's consciousness, in its physically, materially, or, concretely, embodied *form* or *version,* during its wakeful state, takes part, willy-nilly, perforce, without a choice, whether desired or not, or, every which way, each and every day, *fully, completely, all out,* or, *in all respects,* i.e., *to the fullest extent,* is floating, wafting, or levitating, plus whirling, twirling or spiralling nonstop, willy-nilly, perforce, without a choice, whether desired or not, or, every which way, each and every day, and has been doing so from the beginning of *cosmic time* and will continue to do so till the end of *cosmic time.*

What has been said above can be put in another way.

The physical, material, or concrete cosmos, world, or, universe, which human being's consciousness in its physically, materially, or, concretely, embodied *form* or *version,* during its wakeful state, perceives, and experiences and in which it takes part *fully, completely, all out,* or, *in all respects,* i.e., *to the fullest extent,* during its wakeful state, consists of countless physical, material, or concrete moons, planets, stars, black holes, galaxies, and the like.

Thus, all these physical, material, or concrete moons, planets, stars, black holes, galaxies, and the like, which human being's consciousness in its physically, materially, or, concretely, embodied *form* or *version* during its wakeful state, perceives, and experiences, and about which it wonders or speculates, and which it explores or investigates with the help of its own invented and created, such *scientific gadgets, implements,* or, tools as telescopes, exploratory rockets, robotic probes, satellites, and the like, on one hand, and with the help of such direct *human interventions,* as human space flights, human spacewalks, automatic planetary landings, space shuttles, space stations, manned space flights, cosmonauts, astronauts, and the like, on the other, in its wakeful state, are all floating, wafting, or levitating, plus whirling, twirling or spiralling nonstop inside the ubiquitous and infinite field of consciousness called Cosmic Space or Brahmandic Aakash which is the unusual or uncommon, 3-D or three-dimensional form or version or, the unusual or uncommon, expanded, distended, dilated, or inflated form or version of the usually or customarily dimensionless form or version, or, the usually or customarily unexpanded, undistended, undilated, or uninflated form or version of the incredible *consciousness* of God or Brahman i.e., Daydreamer, Oneiricker, Author, Maker, Creator, Progenitor, Primogenitor, Begetter of the physical, material, or concrete cosmos, world, or, universe.

Incidentally, planet earth which is the abode or home of the physically, materially, or, concretely, embodied human consciousness, is also floating, wafting, or levitating, plus whirling, twirling or spiralling nonstop inside the ubiquitous and infinite field of consciousness called Cosmic Space or Brahmandic Aakash which is the unusual or uncommon, 3-D or three-dimensional form or version or, the unusual or uncommon, expanded, distended, dilated, or inflated form or version of the usually or customarily dimensionless form or version, or, the usually or customarily unexpanded, undistended, undilated, or uninflated form or version of the incredible *consciousness* of God or Brahman i.e., Daydreamer, Oneiricker, Author, Maker, Creator, Progenitor, Primogenitor, or Begetter of the physical, material or concrete cosmos, world, or, universe.

Taking Cosmic Space or Brahmandic Aakash as a guide, or model, or, as an archetype, prototype, paradigm, or precedent, or, better still, as a benchmark, or template, the unusual or uncommon, 3-D or three-dimensional form or version or, the unusual or uncommon, expanded, distended, dilated, or inflated form or version of the *subjectively* bodiless, *twin, clone, incarnation, or avatar* of human being's consciousness during latter's dream sleep state, can be said as being very much alike, akin or similar to a *mini, miniature, dwarf, diminutive, small-*

scale, or, scaled-down form or version of Cosmic Space or Brahmandic Akash which exists or abides inside the *subjectively* bodiless, *twin, clone, incarnation, or avatar* of human being's consciousness during latter's dream sleep state.

And inside this *mini, miniature, dwarf, diminutive, small-scale, or, scaled-down* form or version of cosmic space or brahmandic aakash, which obtains, occurs, abides, or, exists, inside the unusual or uncommon, 3-D or three-dimensional form or version or, inside the unusual or uncommon, expanded, distended, dilated, or inflated form or version of the *subjectively* bodiless, *twin, clone, incarnation, or avatar* of human being's consciousness, each and every day, during latter's dream sleep state, God, Brahman, Daydreamer, Oneiricker, Author, Maker, Creator, Progenitor, Primogenitor, or Begetter of the physical, material or concrete cosmos, world, or, universe, that is to say, Cosmic Space or Brahmandic Aakash, which is observed, seen, visualised or watched by human being's consciousness in its wakeful state with the aid of eyes of its physical, material or concrete body, creates or brings into existence a *mini, miniature, dwarf, diminutive, small-scale, or, scaled-down* form or version of the *physical, material, substantial, objecting, or, concrete*-looking, resembling, imitating, or, mimicking, *dreamal,* or *phantasmal* cosmos, world or universe which is very much akin, alike, or, similar in essence, quintessence, or, nature, to the physical, material, substantial, objecting, or, concrete cosmos, world or universe, which human being's consciousness, in its embodied form or version during its wakeful state, observes, sees, visualises or watches, plus, feels, palpates, touches, tastes, hears, and smells, that is to say, perceives and experiences, *fully, completely, all out,* or, *in all respects,* i.e., *to the fullest extent,* and in which it participates, or, takes part *fully, completely, all out,* or, *in all respects,* i.e., *to the fullest extent,* each and everyday, during its wakeful state.

Incidentally, one must underscore or remind oneself anew that God, Brahman, Daydreamer, Oneiricker, Author, Maker, Creator, Progenitor, Primogenitor, or Begetter of the *physical, material or concrete* cosmos, world, or, universe, (the *physical, material or concrete* cosmos, world, or, universe, which human being's consciousness in its physically, materially, or, concretely, embodied *form* or *version,* during its wakeful state, each and every day, perceives and experiences plus in which it takes part, during its wakeful state, each and every day), is no other individual, existence, actuality, *reality,* entity, *truth,* or, *being,* than the *ubiquitous* and *infinite* field of consciousness called Cosmic Space or Brahmandic Aakash, which is the unusual or uncommon, 3-D or three-dimensional form or version or, the unusual or uncommon, expanded, distended, dilated, or inflated form or version of the usually or customarily dimensionless form or version or, the usually or customarily unexpanded, undistended, undilated, or uninflated form or version of the *consciousness* of God, Brahman, daydreamer, oneiricker, Author, Maker, Creator, Progenitor, Primogenitor, or, Begetter, of the *physical, material,* or, *concrete cosmos, world,* or, *universe.*

To state somewhat differently.

A *physical, material, substantial, objective,* or, *concrete*-looking, resembling, imitating, or, mimicking, *mini, miniature, dwarf, diminutive, small-scale, or, scaled-down* SIZE of dreamal or phantasmal cosmos, world or universe, exists, obtains, or occurs, inside the unusual or uncommon, 3-D or three-dimensional form or version or, inside the unusual or uncommon, expanded, distended, dilated, or inflated form or version, of the *subjectively* bodiless, *twin, clone, incarnation, or avatar* of human being's consciousness, each and every day, during latter's dream sleep state.

What has been said above can be put in another way.
A *physical, material, substantial, objective,* or, *concrete*-looking, resembling, imitating, or, mimicking, *mini, miniature, dwarf, diminutive, small-scale, or, scaled-down* model or version, of a cosmos, world or universe,

which is a mere dream, dreamry, imagery, or fantasy and nothing else, or, which is one hundred percent dreamal, phantasmal, consciousnessbal, awarenessbal, or, sentiencel in quintessence, essence, or nature and nothing else, exists, obtains, or occurs, inside the unusual or uncommon, 3-D or three-dimensional form or version or, inside the unusual or uncommon, expanded, distended, dilated, or inflated form or version, of the *subjectively* bodiless, *twin, clone, incarnation, or avatar* of human being's consciousness, each and every day, during latter's dream sleep state.

And this *physical, material, substantial, objective, or, concrete*-looking, resembling, imitating, or, mimicking, mini, miniature, *dwarf, diminutive, small-scale, or, scaled-down* SIZE of dreamal or phantasmal cosmos, world, or, universe, which exists, obtains, or occurs, or, which appears and makes its presence felt, inside the unusual or uncommon, 3-D or three-dimensional *form* or *version,* or, inside the unusual or uncommon, expanded, distended, dilated, or inflated *form* or *version* of the *subjectively* bodiless, *twin, clone, incarnation, or avatar* of human being's consciousness, each and every day, during latter's dream sleep state, is created, generated, brought about, given rise to, set off, or, brought into existence inside the unusual or uncommon, 3-D or three-dimensional *form* or *version,* or, inside the unusual or uncommon, expanded, distended, dilated, or inflated *form* or *version* of the *subjectively* bodiless, *twin, clone, incarnation, or avatar* of human being's consciousness, each and every day, during latter's dream sleep state, by no less an individual, existence, actuality, *reality,* entity, *truth,* or, *being,* than God or Brahman aka the ubiquitous and infinite field of consciousness called Cosmic Space or Brahmandic Aakash who is the Daydreamer, Oneiricker, Author, Maker, Creator, Progenitor, Primogenitor, or Begetter of the *physical, material,* or, *concrete cosmos, world, or, universe* which human being's consciousness in its physically, materially, or, concretely, embodied *form* or *version,* during its wakeful state, willy-nilly, perforce, without a choice, whether desired or not, or, every which way, each and every day, confronts, meets, or, faces, and then, willy-nilly, perforce, without a choice, whether desired or not, or, every which way, each and every day, observes, sees, visualises, watches, feels, palpates, touches, tastes, hears, and smells, that is to say, perceives and experiences, *fully, completely, all out,* or, *in all respects,* i.e., *to the fullest extent,* plus, in which it willy-nilly, perforce, without a choice, whether desired or not, or, every which way, each and every day, participates or takes part, *fully, completely, all out,* or *in all respects,* i.e., *to the fullest extent,* through the *agency, medium,* or, *instrumentality* of its physical, material, or, concrete body, on one hand, and through the *agency, medium,* or, *instrumentality* of the *five* senses or, *five* sense organs of its physical, material, or, concrete body, on the other.

To paraphrase.

And this *physical, material, substantial, objective, or, concrete*-looking, resembling, imitating, or, mimicking, mini, miniature, *dwarf, diminutive, small-scale, or, scaled-down* model or version of a cosmos, world or universe, which is a mere dream, dreamry, imagery, or fantasy and nothing else, or, which is one hundred percent dreamal, phantasmal, consciousnessbal, awarenessbal, or, sentiencel in quintessence, essence, or nature and nothing else, and which exists, obtains, or occurs, or, which appears and makes its presence felt, inside the unusual or uncommon, 3-D or three-dimensional *form* or *version,* or, inside the unusual or uncommon, expanded, distended, dilated, or inflated *form* or *version* of the *subjectively* bodiless, *twin, clone, incarnation, or avatar* of human being's consciousness, each and every day, during latter's dream sleep state, is created, generated, brought about, given rise to, set off, or, brought into existence inside the unusual or uncommon, 3-D or three-dimensional *form* or *version,* or, inside the unusual or uncommon, expanded, distended, dilated, or inflated *form* or *version* of the *subjectively* bodiless, *twin, clone, incarnation, or avatar* of human being's consciousness, each and every day, during latter's dream sleep state, by no less an individual, existence, actuality, *reality,* entity, *truth,* or, *being,* than God or Brahman aka the ubiquitous and infinite field of consciousness called Cosmic Space or Brahmandic Aakash who is the Daydreamer, Oneiricker, Author, Maker, Creator, Progenitor, Primogenitor, or Begetter of the *physical, material,* or, *concrete cosmos, world, or, universe* which human being's consciousness in its physically, materially, or, concretely, embodied *form* or *version,* during its wakeful state, willy-nilly, perforce, without a choice, whether desired or not, or, every which way, each and every day, confronts, meets, or, faces, and then, willy-nilly, perforce, without a choice, whether desired or not, or, every which way, each and every day, observes, sees, visualises, watches, feels, palpates, touches, tastes, hears, and smells, that is to say, perceives and experiences, *fully, completely, all out,* or, *in all respects,* i.e., *to the fullest extent,* plus, in which it willy-nilly, perforce, without a choice, whether desired or not, or, every which way, each and every day, participates or takes part, *fully, completely, all out,* or *in all respects,* i.e., *to the fullest extent,* through the *agency, medium,* or, *instrumentality* of its physical, material, or, concrete body, on one hand, and through the *agency, medium,* or, *instrumentality* of the *five* senses or, *five* sense organs of its physical, material, or, concrete body, on the other.

To repeat.

And this *physical, material, substantial, objective, or, concrete*-looking, resembling, imitating, or, mimicking, mini, miniature, *dwarf, diminutive, small-scale, or, scaled-down* SIZE of dreamal or phantasmal cosmos, world or universe, which obtains, occurs, or, exists, or, which appears and makes its presence felt, inside the unusual or uncommon, 3-D or three-dimensional form or version or, inside the unusual or uncommon, expanded, distended, dilated, or inflated form or version of the *subjectively* bodiless, *twin, clone, incarnation, or avatar* of human being's consciousness, each and every day, during latter's dream sleep state, is created, generated, brought about, given rise to, set off, or, brought into existence, inside the unusual or uncommon, 3-D or three-dimensional *form* or *version,* or, inside the unusual or uncommon, expanded, distended, dilated, or inflated *form* or *version* of the *subjectively* bodiless, *twin, clone, incarnation, or avatar* of human being's consciousness, each and every day, during latter's dream sleep state, willy-nilly, perforce, without a choice, whether desired or not, or, every which way, each and every day, by no less an individual, existence, actuality, *reality,* entity, *truth,* or, *being,* than God or Brahman aka the ubiquitous and infinite field of consciousness called Cosmic Space or Brahmandic Aakash who is the Daydreamer, Oneiricker, Author, Maker, Creator, Progenitor, Primogenitor, or Begetter of the *physical, material,* or, *concrete cosmos, world, or, universe* which human being's consciousness in its physically, materially, or, concretely, embodied *form* or *version,* during its wakeful state, willy-nilly, perforce, without a choice, whether desired or not, or, every which way, each and every day, confronts, meets, or, faces, and then, willy-nilly, perforce, without a choice, whether desired or not, or, every which way, each and every day, observes, sees, visualises, watches, feels, palpates, touches, tastes, hears, and smells, that is to say, perceives and experiences, *fully, completely, all out,* or, *in all respects,* i.e., *to the fullest extent,* plus, in which it willy-nilly, perforce, without a choice, whether desired or not, or, every which way, each and every day, participates or takes part, *fully, completely, all out,* or *in all respects,* i.e., *to the fullest extent,* through the *agency, medium,* or, *instrumentality* of its physical, material, or, concrete body, on one hand, and through the *agency, medium,* or, *instrumentality* of the *five* senses or, *five* sense organs of its physical, material, or, concrete body, on the other.

The whole state of affairs, described above, is the actuality, reality, fact, truth, the real world, real life, existence, or living, because Cosmic Space or Brahmandic Aakash, which human being's consciousness, in its physically, materially, or, substantially, *embodied form or version,* during its wakeful state, willy-nilly, perforce, without a choice, whether desired or not, or, every which way, each and every day, confronts, meets, or, faces plus observes, sees, visualises, or, watches with the aid or the help of the eyes of its physical, material, or, concrete *body,* and about which it daily wonders, conjectures, cudgels its brains about, thinks about, deliberates about, speculates about, puzzles about, is curious about, is inquisitive about, meditates on, or reflects on, with regards to or, vis-a-vis its basic, bottom line, quintessential, fundamental, ultimate, supreme, or absolute nature, is nothing but the unusual or uncommon, 3-D or three-dimensional form or version or, the unusual or uncommon, expanded, distended, dilated, or inflated form or version of the usually, or, customarily, dimensionless form or version or, the usually or, customarily, unexpanded, undistended, undilated, or uninflated form or version of the *consciousness* of God or Brahman or, the consciousness of the Daydreamer, Oneiricker, Author, Maker, Creator, Progenitor, Primogenitor, or Begetter of the *physical, material,* or, *concrete* cosmos, world, or, universe.

The *physical, material, substantial, objective, or, concrete*-looking, resembling, imitating, or, mimicking, mini, miniature, *dwarf, diminutive, small-scale, or, scaled-down* model or version of a cosmos, world or universe, which is a mere dream, dreamry, imagery, or fantasy and nothing else, or, which is one hundred percent dreamal, phantasmal, consciousnessbal, awarenessbal, or, sentiencel in quintessence, or nature, and nothing else, and which exists, obtains, or occurs, or, which appears and makes its presence felt, inside the unusual or uncommon, 3-D or three-dimensional *form* or *version,* or, inside the unusual or uncommon, expanded, distended, dilated, or inflated *form* or *version* of the *subjectively* bodiless, *twin, clone, incarnation, or avatar* of human being's consciousness, during latter's dream sleep state, willy-nilly, perforce, without a choice, whether desired or not, or, every which way, each and every day, and which the *objectively* embodied *twin, clone, incarnation* or *avatar* of human being's consciousness, or, absolutely to the point, which the *consciousnessbally* embodied, *twin, clone, incarnation* or *avatar* of human being's consciousness, during latter's dream sleep state, willy-nilly, perforce, without a choice, whether desired or not, or, every which way, each and every day, confronts, meets, or, faces plus observes, sees, visualises, watches, feels, palpates, touches, tastes, hears, and smells, that is to say, perceives and experiences, *fully, completely, all out,* or, *in all respects,* i.e., *to the fullest extent,* and, in which, it willy-nilly, perforce, without a choice, whether desired or not, or, every which way, each and every day, participates, or, takes part *fully, completely, all out,* or, *in all respects,* i.e., *to the fullest extent,* with the aid or assistance of its consciousnessbal-substance made or consciousnessbal-matter made *body* & this body's five consciousnessbal-substance made or consciousnessbal-matter made *sense organs* during *dream sleep state* of human being's consciousness, is created, generated, brought about, set off, given rise to, or, brought into existence, inside the unusual or uncommon, 3-D or three-dimensional *form* or *version,* or, inside the unusual or uncommon, expanded, distended, dilated, or inflated *form* or *version* of the *subjectively* bodiless, *twin, clone, incarnation, or avatar* of human being's consciousness during latter's dream sleep state, willy-nilly, perforce, without a choice, whether desired or not, or, every which way, each and every day, by no less an individual, actuality, *reality,* entity, *truth,* existence, or, *being,* than the ubiquitous and infinite field of consciousness called Cosmic Space or Brahmandic Aakash aka God, Brahman, Daydreamer, Oneiricker, Author, Maker, Creator, Progenitor, Primogenitor, or Begetter of the *physical, material,* or, *concrete cosmos, world, or, universe* of the wakeful state with the sole purpose of apprising, advising, enlightening, or, informing, human being's consciousness during its wakeful state that the *physical, material, substantial, objective,* or, *concrete* cosmos, world, or, universe, which human being's consciousness in its embodied form or version during its wakeful state, willy-nilly, perforce, without a choice, whether desired or not, or, every which way, each and every day, observes, sees, visualises, watches, feels, palpates, touches, tastes, hears, &, smells, that is to say, perceives and experiences, *fully, completely, all out,* or, *in all respects,* i.e., *to the fullest extent,* and, in which, it willy-nilly, perforce, without a choice, whether desired or not, or, every which way, each and every day, participates, or, takes part *fully, completely, all out,* or, *in all respects,* i.e., *to the fullest extent,* is also of *dreamal* or *phantasmal* nature just as is the *dreamal* or *phantasmal* cosmos, world or universe which the *objectively* embodied *twin, clone, incarnation* or *avatar* of human being's consciousness, or, absolutely to the point, which the *consciousnessbally* embodied, *twin, clone, incarnation* or *avatar* of human being's consciousness, during latter's dream sleep state, willy-nilly, perforce, without a choice, whether desired or not, or, every which way, each and every day, observes, sees, visualises, watches, feels, palpates, touches, tastes, hears, and smells, that is to say, perceives and experiences, *fully, completely, all out,* or, *in all respects,* i.e., *to the fullest extent,* and, in which, it willy-nilly, perforce, without a choice, whether desired or not, or, every which way, each and every day, participates, or, takes part *fully, completely, all out,* or, *in all respects,* i.e., *to the fullest extent* during human being's dream sleep state.

To repeat.

The *physical, material, substantial, objective, or, concrete*-looking, resembling, imitating, or, mimicking, mini, miniature, *dwarf, diminutive, small-scale, or, scaled-down* model or version of a cosmos, world or universe, which is a mere dream, dreamry, imagery, or fantasy and nothing else, or, which is one hundred percent dreamal, phantasmal, consciousnessbal, awarenessbal, or, sentiencel in quintessence, or nature, and nothing else, and which exists, obtains, or occurs, or, which appears and makes its presence felt, inside the unusual or uncommon, 3-D or three-dimensional *form* or *version,* or, inside the unusual or uncommon, expanded, distended, dilated, or inflated *form* or *version* of the *subjectively* bodiless, *twin, clone, incarnation, or avatar* of human being's consciousness, during latter's dream sleep state, willy-nilly, perforce, without a choice, whether desired or not, or, every which way, each and every day, and, which the *objectively* embodied *twin, clone, incarnation* or *avatar* of human being's consciousness, or, absolutely to the point, which the *consciousnessbally* embodied, *twin, clone, incarnation* or *avatar* of human being's consciousness, during latter's dream sleep state, willy-nilly, perforce, without a choice, whether desired or not, or, every which way, each and every day, confronts, meets, or, faces plus observes, sees, visualises, watches, feels, palpates, touches, tastes, hears, and, smells, that is to say, perceives and experiences, *fully, completely, all out,* or, *in all respects,* i.e., *to the fullest extent,* and, in which, it willy-nilly, perforce, without a choice, whether desired or not, or, every which way, each and every day, participates, or, takes part *fully, completely, all out,* or, *in all respects,* i.e., *to the fullest extent,* with the aid or assistance of its consciousnessbal-substance made or consciousnessbal-matter made *body* and this body's five consciousnessbal-substance made or consciousnessbal-matter made *sense organs* during *dream sleep state* of human being's consciousness, is created, generated, brought about, set off, given rise to, or, brought into existence, inside the unusual or uncommon, 3-D or three-dimensional *form* or *version,* or, inside the unusual or uncommon, expanded, distended, dilated, or inflated *form* or *version* of the *subjectively* bodiless, *twin, clone, incarnation, or avatar* of human being's consciousness during latter's dream sleep state, willy-nilly, perforce, without a choice, whether desired or not, or, every which way, each and every day, by no less an individual, actuality, *reality,* entity, *truth,* existence, or, *being,* than the ubiquitous and infinite field of consciousness called Cosmic Space or Brahmandic Aakash aka God, Brahman, Daydreamer, Oneiricker, Author, Maker, Creator, Progenitor, Primogenitor, or Begetter of the *physical, material,* or, *concrete cosmos, world, or, universe* of the wakeful state with the sole purpose of apprising, advising, enlightening, or, informing, human being's consciousness during its wakeful state that the *physical, material, substantial, objective,* or, *concrete* cosmos, world, or, universe, which human being's consciousness in its embodied form or version during its wakeful state, willy-nilly, perforce, without a choice, whether desired or not, or, every which way, each and every day, observes, sees, visualises, watches, feels, palpates, touches, tastes, hears, and smells, that is to say, perceives and experiences, *fully, completely, all out,* or, *in all respects,* i.e., *to the fullest extent,* and, in which, it willy-nilly, perforce, without a choice, whether desired or not, or, every which way, each and every day, participates, or, takes part *fully, completely, all out,* or, *in all respects,* i.e., *to the fullest extent,* is also of *dreamal* or *phantasmal* nature just as is the *dreamal* or *phantasmal* cosmos, world or universe which the *objectively* embodied *twin, clone, incarnation* or *avatar* of human being's consciousness, or, absolutely to the point, which the *consciousnessbally* embodied, *twin, clone, incarnation* or *avatar* of human being's consciousness, during latter's dream sleep state, willy-nilly, perforce, without a choice, whether desired or not, or, every which way, each and every day, observes, sees, visualises, watches, feels, palpates, touches, tastes, hears, and smells, that is to say, perceives and experiences, *fully, completely, all out,* or, *in all respects,* i.e., *to the fullest extent,* and, in which, it willy-nilly, perforce, without a choice, whether desired or not, or, every which way, each and every day, participates, or, takes part *fully, completely, all out,* or, *in all respects,* i.e., *to the fullest extent* during human being's dream sleep state.

Both kinds of cosmoses, worlds or universes, namely, "the first one", which is correctly labeled as a mere dream, dreamry, imagery, or fantasy, or, as a mere *dreamal,* or, *phantasmal* cosmos, world, or, universe and which exists, occurs, or obtains inside the unusual or uncommon, 3-D or three-dimensional form or version or, inside the unusual or uncommon, expanded, distended, dilated, or inflated form or version of the *subjectively* bodiless, *twin, clone, incarnation, or avatar* of human being's consciousness during latter's dream sleep state and which is observed, seen, visualised, or watched, plus, experienced to the limited extent, of merely observing, seeing, visualising, or watching, by this bodiless, *twin, clone, incarnation, or avatar* of human being's consciousness during its dream sleep state, on one hand, and "the second one" which is incorrectly labelled as one hundred percent actual, factual, real, genuine, or, authentic, and is tagged, or, titled, as *physical, material, or, concrete* cosmos, world, or, universe by human being's consciousness during its wakeful state and which exists, occurs, or obtains, inside the *omnipresent, all-present, all-pervading, spreading through or into every, universal, ubiquitous,* or, *infinite* field of consciousness called Cosmic Space or Brahmandic Aakash, during the wakeful state of human being's consciousness, on the other, are created, brought about, set off, given rise to, or brought into being, by the latter only and no one else, that is to say, are created, brought about, set off, given rise to, or, brought into being, by the *omnipresent, all-present, all-pervading, spreading through or into every, universal, ubiquitous,* or, *infinite* field of consciousness called Cosmic Space or Brahmandic Aakash only and no one else.

"The second", *dreamal,* or, *phantasmal* cosmos, world, or, universe, mentioned above, namely, the one, which human being's consciousness in its embodied form or version during its wakeful state, willy-nilly, perforce, without a choice, whether desired or not, or, every which way, each and every day, observes, sees, visualises, watches, feels, palpates, touches, tastes, hears, and smells, that is to say, perceives and experiences, *fully, completely, all out,* or, *in all respects,* i.e., *to the fullest extent,* and, in which, it willy-nilly, perforce, without a choice, whether desired or not, or, every which way, each and every day, participates, or, takes part *fully, completely, all out,* or, *in all respects,* i.e., *to the fullest extent,* that is to say, i.e., in other words, "the second " *dreamal* or *phantasmal* cosmos, world or universe, mentioned above, which by sight and tactile sense, by eyeball and tangibility, by vision and touch, by look and texture, or, by eyeshot and feel, appears or seems as if it is one hundred percent *physical, material, substantial,* or, *concrete* in nature and which exists, occurs, or obtains inside the unusual or uncommon, 3-D or three-dimensional form or version or, inside the unusual or uncommon, expanded, distended, dilated, or inflated form or version of the consciousness of God or Brahman viz., which exists, occurs, or obtains inside the *omnipresent, all-present, all-pervading, spreading through or into every, universal, ubiquitous,* or, *infinite* field of consciousness called Cosmic Space or Brahmandic Aakash, during the wakeful state of human being's consciousness, when human being's consciousness is in existence in its *physical, material, substantial,* or, *concrete* looking and feeling *embodied* form or version, is created, given rise to, brought about, set off, triggered, begotten, or, begot by the consciousness of God or Brahman viz., is created, brought about, set off, triggered, begotten, or, begot by the *omnipresent, all-present, all-pervading, spreading through or into everything, universal, ubiquitous,* or, *infinite* field of consciousness called Cosmic Space or Brahmandic Aakash, inside Itself through the activity of daydreaming or, oneiricking on its part only and nothing else, with the sole aim of amusing, entertaining, or, regaling itself, or, with the sole aim of pleasing, delighting, or enthralling Itself.

What has been said above, can be put slightly differently.

"The second" *dreamal* or *phantasmal* cosmos, world or universe, mentioned above, which by sight and tactile sense, by eyeball and tangibility, by vision and touch, by look and texture, or, by eyeshot and feel, appears or seems as if it is one hundred percent *physical, material, substantial,* or, *concrete* in nature and which exists, occurs, or obtains inside the unusual or uncommon, 3-D or three-dimensional form or version or, inside the unusual or uncommon, expanded, distended, dilated, or inflated form or version of the consciousness of God or Brahman viz., which exists, occurs, or obtains inside the *omnipresent, all-present, all-pervading, spreading through or into everything, universal, ubiquitous,* or, *infinite* field of consciousness called Cosmic Space or Brahmandic Aakash, during the wakeful state of human being's consciousness when human being's consciousness is in existence in its *physical, material, substantial,* or, *concrete* looking and feeling *embodied* form or version, is created, given rise to, brought about, set off, triggered, begotten, or, begot by the consciousness of God or Brahman viz., is created, brought about, set off, triggered, begotten, or, begot by the *omnipresent, all-present, all-pervading, spreading through or into everything, universal, ubiquitous,* or, *infinite* field of consciousness called Cosmic Space or Brahmandic Aakash, inside Itself through the activity of daydreaming or, oneiricking on its part only and nothing else, with the sole aim of amusing, entertaining, or, regaling itself, or, with the sole aim of pleasing, delighting, or enthralling Itself.

To rephrase.

"The second", *dreamal* or *phantasmal* cosmos, world or universe, mentioned above, which is *physical, material, substantial,* or, *concrete* looking and feeling, and, which exists, occurs, obtains or is in existence inside the unusual or uncommon, 3-D or three-dimensional form or version or, inside the unusual or uncommon, expanded, distended, dilated, or inflated form or version of the consciousness of God or Brahman i.e., which exists, occurs, obtains, or, is in existence, inside the *omnipresent, all-present, all-pervading, spreading through or into everything, universal, ubiquitous,* or, *infinite* field of consciousness called Cosmic Space or Brahmandic Aakash, during the wakeful state of human being's consciousness when the human being's consciousness exists, occurs, obtains, or, is in existence, in its *physical, material, substantial,* or, *concrete* looking and feeling, *embodied* form or version, is created, brought about, set off, triggered, begotten, or, begot by the consciousness of God or Brahman viz., is created, brought about, set off, triggered, begotten, or, begot by the *omnipresent, all-present, all-pervading, spreading through or into everything, universal, ubiquitous,* or, *infinite* field of consciousness called Cosmic Space or Brahmandic Aakash inside Itself through the activity of daydreaming or, oneiricking on its part only and nothing else, with the sole aim of amusing, entertaining, or, regaling itself, or, with the sole aim of pleasing, delighting, or enthralling Itself.

Therefore, the above mentioned, "second", *dreamal* or *phantasmal* cosmos, world or universe, which by sight and tactile sense, by eyeball and tangibility, by vision and touch, by look and texture, or, by eyeshot and feel, appears or seems as if it is one hundred percent *physical, material, substantial,* or, *concrete* in nature and which exists, occurs, or obtains inside the unusual or uncommon, 3-D or three-dimensional form or version or, inside the unusual or uncommon, expanded, distended, dilated, or inflated form or version of the consciousness of God or Brahman viz., which exists, occurs, or obtains inside the *omnipresent, all-present, all-pervading, spreading through or into everything, universal, ubiquitous,* or, *infinite* Field of Consciousness called Cosmic Space or Brahmandic Aakash during the wakeful state of human being's consciousness when the human being's consciousness is in existence in its *physical, material, substantial,* or, *concrete* looking and feeling *embodied* form or version, and, which has been created, given rise to, brought about, set off, triggered, begotten, or, begot by the consciousness of God or Brahman viz., and which has been created, given rise to, brought about, set of, triggered, begotten, or, begot by the *omnipresent, all-present, all-pervading, spreading through or into everything, universal, ubiquitous,* or, *infinite* Field of Consciousness called Cosmic Space or Brahmandic Aakash, inside Itself through the activity of daydreaming or, oneiricking on its part only, and nothing else, with the sole aim of amusing, entertaining, or, regaling itself, or, with the sole aim of pleasing, delighting, or enthralling Itself, is called an exquisite, beautiful, excellent, outstanding, incomparable, matchless, perfect, plus, finely detailed, finely honed, and, finely tuned "LEELA" of the consciousness of God or Brahman viz., is called an exquisite, beautiful, excellent, outstanding, incomparable, matchless, perfect, plus, finely detailed, finely honed, and, finely tuned "LEELA" of the *omnipresent, all-present, all-pervading, spreading through or into everything, universal, ubiquitous,* or, *infinite* Field of Consciousness called Cosmic Space or Brahmandic Aakash in the realm or precinct of Adwait-Vedanta.

To repeat.

Therefore, the above mentioned, "second", *dreamal* or *phantasmal* cosmos, world or universe, which by sight and tactile sense, by eyeball and tangibility, by vision and touch, by look and texture, or, by eyeshot and feel, appears or seems as if it is one hundred percent *physical, material, substantial,* or, *concrete* in nature and which exists, occurs, or obtains inside the unusual or uncommon, 3-D or three-dimensional form or version or, inside the unusual or uncommon, expanded, distended, dilated, or inflated form or version of the consciousness of God or Brahman viz., which exists, occurs, or obtains inside the *omnipresent, all-present, all-pervading, spreading through or into everything, universal, ubiquitous,* or, *infinite* Field of Consciousness called Cosmic Space or Brahmandic Aakash during the wakeful state of human being's consciousness, when human being's consciousness is in existence in its *physical, material, substantial,* or, *concrete* looking and feeling *embodied* form or version, and, which has been created, given rise to, brought about, set off, triggered, begotten, or, begot by the consciousness of God or Brahman viz., and which has been created, given rise to, brought about, set of, triggered, begotten, or, begot by the *omnipresent, all-present, all-pervading, spreading through or into everything, universal, ubiquitous,* or, *infinite* Field of Consciousness called Cosmic Space or Brahmandic Aakash, inside Itself through the activity of daydreaming or, oneiricking on its part only, and nothing else, with the sole aim of amusing, entertaining, or, regaling itself, or, with the sole aim of pleasing, delighting, or enthralling Itself, is called an exquisite, beautiful, excellent, outstanding, incomparable, matchless, perfect, plus, finely detailed, finely honed, and, finely tuned, "LEELA" of the consciousness of God or Brahman viz., is called an exquisite, beautiful, excellent, finely detailed, outstanding, incomparable, matchless, perfect, plus, finely detailed, finely honed, and, finely tuned, "LEELA" of the *omnipresent, all-present, all-pervading, spreading through or into everything, universal, ubiquitous,* or, *infinite F*ield of Consciousness called Cosmic Space or Brahmandic Aakash, in the realm or precinct of Adwait-Vedanta.

Before explaining what the Sanskrit word "LEELA" means, let one emphasise anew or underscore that, this second *physical, material, substantial,* or, *concrete* looking and feeling *dreamal* or *phantasmal* cosmos, world or universe which is called an exquisite, beautiful, excellent, outstanding, incomparable, matchless, perfect, plus, finely detailed, finely honed, and, finely tuned, "LEELA" of the consciousness of God or Brahman viz., which is called an exquisite, beautiful, excellent, outstanding, incomparable, matchless, perfect, plus, finely detailed, finely honed, and finely tuned, "LEELA" of the *omnipresent, all-present, all-pervading, spreading through or into everything, universal, ubiquitous,* or, *infinite* Field of Consciousness called Cosmic Space or Brahmandic Aakash in the realm or precinct of Adwait-Vedanta, is created, given rise to, brought about, set off, triggered, begotten, or, begot by the consciousness of God or Brahman viz., is created, given rise to, brought about, triggered, begotten, or, begot by the *omnipresent, all-present, all-pervading, spreading through or into everything, universal, ubiquitous,* or, *infinite* Field of Consciousness called Cosmic Space or Brahmandic Aakash through the activity of daydreaming or oneiricking on its part only and nothing else.

And this second *physical, material, substantial,* or, *concrete* looking and feeling *dreamal* or *phantasmal* cosmos, world or universe, which is called an exquisite, beautiful, excellent, outstanding, incompatible, matchless, perfect, plus, finely detailed, finely honed, and finely tuned, "LEELA" of the consciousness of God or Brahman viz., which is called an exquisite, beautiful, excellent, outstanding, incompatible, matchless, perfect, plus, finely detailed, finely honed, and finely tuned, "LEELA" of the *omnipresent, all-present, all-pervading, spreading through or into everything, universal, ubiquitous,* or, *infinite* Field of Consciousness called Cosmic Space or Brahmandic Aakash in the realm or precinct of Adwait-Vedanta, is that incredible or extraordinary, *dreamal* or *phantasmal* cosmos, world or universe which human being's consciousness in its *physical, material, substantial,* or, *concrete* looking and feeling *embodied* form or version during its wakeful state, willy-nilly, perforce, without a choice, whether desired or not, or, every which way, each and every day, confronts, meets, or, faces plus observes, sees, visualizes, watches, feels, palpates, touches, tastes, hears, and smells, that is to say, perceives and experiences, *fully, completely, all out,* or, *in all respects,* i.e., *to the fullest extent,* and, in which, it participates, or, takes part *fully, completely, all out,* or, *in all respects,* i.e., *to the fullest extent,* willy-nilly, perforce, without a choice, whether desired or not, or, every which way, each and every day.

Here one must also remind oneself anew, or, underscore that the *omnipresent, all-present, all-pervading, spreading through or into everything, universal, ubiquitous,* or, *infinite* Field of Consciousness called Cosmic Space or Brahmandic Aakash, which human being's consciousness in its *physical, material, or, concrete* looking and feeling *embodied* form or version during its wakeful state, willy-nilly, perforce, without a choice, whether desired or not, or, every which way, each and every day, confronts, meets, or, faces plus observes, sees, visualises, or watches with the help of eyes of its *physical, material, or, concrete* looking and feeling, *dreamal* or *phantasmal* body, is the breathtaking, dazzling, amazing, stunning, astounding, astonishing, awe-inspiring, or, staggering, 3-D or three-dimensional form or version or, the expanded, distended, dilated, or, inflated form or version of the dimensionless state or, mode of being or, the original state or mode of being of the consciousness of God or Brahman.

The Sanskrit word, term, or expression "LEELA" means the following :-

"LEELA" (Sanskrit: लीला), like many Sanskrit words, cannot be precisely translated into English, but can be loosely translated as the *noun* "play".

"LEELA" is a way of describing all existences or realities, including the *physical, material, or, concrete* looking and feeling, *dreamal* or *phantasmal* cosmos, world, or, universe, which is perceived and experienced by human being's consciousness during its wakeful state as well as the *physical, material, or, concrete* looking and feeling, *dreamal* or *phantasmal* cosmos, world, or, universe i.e., *dream* or *fantasy,* which is perceived and experienced by human being's consciousness during its dream sleep state, as the outcome of the "creative play" by the *consciousness* of the Divine Absolute i.e., as the outcome of the "creative play" by the *consciousness* of God or Brahman, which is none other than the Absolute and Eternal existence or the Absolute and Eternal reality called the *omnipresent, all-present, all-pervading, spreading through or into everything, universal, ubiquitous,* or, *infinite* Field of Consciousness aka Cosmic Space or Brahmandic Aakash aka the Creator, Maker, or Progenitor of *physical matter,* on one hand, and the Source or Spring of human being's consciousness, on the other.

What has been said above with regards to the Sanskrit world "Leela" can be put in another way.

The concept of "LEELA" within the realm or precinct of

Adwait-Vedanta, is a way of describing all *physical, material, or, concrete* looking and feeling *phenomena,* which are perceived and experienced by human being's consciousness, irrespective of whether these *physical, material, or, concrete* looking and feeling *phenomena* are perceived and experienced by human being's consciousness during its dream sleep state, or, during its wakeful state, as the outcome of the "*creative play*" of the consciousness of the *Divine Absolute* viz., as the outcome of the "creative play" of the consciousness of God or, Brahman who is none other than the Absolute and Eternal existence or reality called the *omnipresent, all-present, all-pervading, spreading through or into everything, universal, ubiquitous,* or, *infinite* Field of Consciousness aka Cosmic Space or Brahmandic Aakash aka the Creator, Maker, or Progenitor of the *physical matter,* on one hand, and the Source or Spring of the human being's consciousness, on the other.

Here, the expression "phenomena" includes all the *physical, material, or, concrete* looking and feeling existences or realities, which are perceived and experienced by human being's consciousness during all its states or, modes of being. That is to say, here the expression "phenomena" includes all the *physical, material, or, concrete* looking and feeling existences or realities, which are perceived and experienced by human being's consciousness during its wakeful state, on one hand, and, during its dream sleep state, on the other.

Here, it will not be inappropriate to remind oneself anew, or, to refresh one's memory anew, or, to underscore that, the consciousness of God or Brahman, or, the consciousness of the Creator, Maker, or, Progenitor of the *physical matter,* as well as the consciousness, who is the Source or Spring of human being's consciousness, in the cosmos, world, or, universe, is none other than the omnipresent, all-present, all-pervading, spreading through or into everything, universal, ubiquitous, or, infinite Field of Consciousness or, Field of Mind, which is called Cosmic Space or Brahmandic Aakash, by human being's consciousness, when human being's consciousness is in its wakeful state, that is to say, when human being's consciousness is in its *physical, material, substantial,* or, *concrete* looking and feeling, *embodied* state, or, mode of being.

The reason why the consciousness of the *Divine Absolute* viz., the consciousness of God or, Brahman, that is to say, the consciousness who is the Creator, Maker, or, Progenitor of the *physical matter,* on one hand, and the Source or Spring of the human being's consciousness in the cosmos, world, or, universe, on the other, namely, the *omnipresent, all-present, all-pervading, spreading through or into everything, universal, ubiquitous,* or, *infinite* Field of consciousness, or, Field of Mind called

Cosmic Space or Brahmandic Aakash, got involved in Its "*creative* play" or, "LEELA", 13.7 billion light years ago viz., before the beginning of Cosmic Time, has been beautifully or splendidly described in the following manner :-

"Before the *beginning* of Cosmic Time, the incredible or unbelievable, dimensionless consciousness of the *Divine Absolute,* that is to say, the incredible or unbelievable, dimensionless consciousness of God or Brahman, or, better still, the incredible or unbelievable, dimensionless consciousness who is the Creator, Maker, or, Progenitor of the *physical matter,* on one hand, and the Source or Spring of the human being's consciousness in the cosmos, world, or, universe, on the other, was totally alone. There was no one to know It, no one to love It. Therefore, it separated a part of Itself from Itself within Itself and out of this part of Itself, It created within Itself a *physical, material, substantial, or concrete* looking and feeling multifarious, multifaceted, or, multiform *dreamal* or *phantasmal* cosmos, world, or, universe of countless *dreamal* or *phantasmal* beings, things, events, or, phenomena through the activity of Its *daydreaming* or *oneiricking* in order that It can love Itself by Itself plus amuse, entertain, regale or enthrall Itself by Itself and nothing else."

What has been said above, can be expressed slightly

differently.

The reason why the consciousness of the *Divine Absolute* viz., of the consciousness of God or, Brahman, namely, the *omnipresent, all-present, all-pervading, spreading through or into everything, universal, ubiquitous,* or, *infinite* Field of consciousness, or, Field of Mind, called Cosmic Space or Brahmandic Aakash aka the Creator, Maker, or Progenitor of the *physical matter* in the cosmos, world, or, universe, on one hand, and the Source or Spring of the human being's consciousness in the cosmos, world, or, universe, on the other, got involved in Its *"creative* play" or, "LEELA", 13.7 billion light years ago viz., before the beginning of Cosmic Time, has been beautifully or splendidly described in the following manner :-

"Before the *beginning* of Cosmic Time, the incredible or unbelievable, dimensionless consciousness of the *Divine Absolute,* that is to say, the incredible or unbelievable, dimensionless consciousness of God or Brahman, or, better still, the incredible or unbelievable, dimensionless consciousness who is the Creator, Maker, or, Progenitor of the *physical matter,* on one hand, and the Source or Spring of the human being's consciousness in the cosmos, world, or, universe, on the other, was totally alone. There was no one to know It, and no one to love It. Therefore, it separated a part of Itself from Itself within Itself and out of this part of Itself, It created within Itself a *physical, material, substantial, or concrete* looking and feeling, multifarious, multifaceted, or, multiform *dreamal* or *phantasmal* cosmos, world, or, universe of countless *dreamal* or *phantasmal* beings, things, events, or, phenomena, through the activity of Its *daydreaming* or *oneiricking,* in order that It can love Itself by Itself plus amuse, entertain, regale or enthrall Itself by Itself and nothing else."

Now let one pick up the original narrative, theme, or thread, or, if one likes, discussion-thread, once again from where one left it earlier. This narrative, theme, or thread, or, discussion-thread was with regards to the first *dreamal* or *phantasmal* cosmos, world or universe, which exists, occurs, or obtains inside the unusual or uncommon, 3-D or three-dimensional form or version or, inside the unusual or uncommon, expanded, distended, dilated, or inflated form or version of the *subjectively* bodiless, *twin, clone, incarnation, or avatar* of human being's consciousness during latter's dream sleep state.

Here, one needs to remind oneself, or, one needs to refresh one's memory, that, there exist, occur, or, obtain, two *dreamal* or *phantasmal* cosmoses, worlds or universes, both of which are willy-nilly, perforce, without a choice, whether desired or not, or, every which way, each and every day encountered, met, or, faced plus observed, seen, visualized, watched, felt, palpated, touched, tasted, heard, and smelt, that is to say, perceived and experienced, *fully, completely, all out,* or, *in all respects,* i.e., *to the fullest extent,* and, both of which are willy-nilly, perforce, without a choice, whether desired or not, or, every which way, each and every day participated in, or, taken part in, *fully, completely, all out,* or, *in all respects,* i.e., *to the fullest extent,* by human being's consciousness each and every day without fail.

The names of these two *dreamal* or *phantasmal* cosmoses, worlds or universes are :- (1) the dream sleep state cosmos, world, or, universe, on one hand, and (2) the wakeful state cosmos, world, or, universe, on the other, both of which materialize without fail each 24 hours everywhere from nowhere during the entire life of human being's consciousness. And both these *dreamal* or *phantasmal* cosmoses, worlds or universes cease to exist, as for as human being's consciousness is concerned, only when the *physical, material,* or *concrete* looking and feeling body, whose presence human being's consciousness feels around itself during its wakeful state *only* , each and every day, finally says "good-bye" i.e., finally "dies" , "demises" , "gives up the ghost" , "gives up its ivy-like hold on the consciousness" , "kicks the bucket" , "relinquishes life" , "relinquishes consciousness" or "goes way of all flesh" and human being's consciousness coalesces back or merges back into the *infinite, boundless, limitless, omnipresent, omniscient, omnipotent, ubiquitous, universal, all-present,* or, *all-pervasive Ocean* of *Consciousness* or *Ocean* of *Mind* or, *Field* of *Consciousness* or *Field* of *Mind* , which is erroneously called, named, labeled, or tagged as Cosmic Space or Brahmandic Aakash by human being's consciousness during its wakeful state due to its nescience but which, in absolute reality , is the unusual or uncommon, 3-D or three-dimensional form or version or, the unusual or uncommon, expanded, distended, dilated, or inflated form or version of the usually, or, customarily, dimensionless form or version or, the usually or, customarily, unexpanded, undistended, undilated, or uninflated form or version of the *consciousness* of God or Brahman or, the consciousness of the Daydreamer, Oneiricker, Author, Maker, Creator, Progenitor, Primogenitor, or Begetter of the *physical, material,* or, *concrete* looking or feeling cosmos, world, or, universe of the wakeful state of human being's consciousness, on one hand, and the *physical, material,* or, *concrete* looking or feeling cosmos, world, or, universe of the dream sleep state of human being's consciousness, on the other, both of which , in absolute reality , are nothing but *dreamal* or *phantasmal* in nature, and created or made by this God, Brahman, Daydreamer, Oneiricker, Author, Maker, Creator, Progenitor, Primogenitor, or Begetter, i.e., this *infinite, boundless, limitless, omnipresent, omniscient, omnipotent, ubiquitous, universal, all-present,* or, *all-pervasive Ocean* of *Consciousness* or *Ocean* of *Mind* or, *Field* of *Consciousness* or *Field* of *Mind* , which is erroneously called, named, labeled, or tagged as Cosmic Space or Brahmandic Aakash by human being's consciousness during its wakeful state due to its nescience.

Now, let one describe once again, the site, location, or, place of residence, existence, presence, or, being of both these *dreamal* or *phantasmal* cosmoses, worlds or universes.

The site, location, or, place of residence, existence, presence, or, being of the first *dreamal* or *phantasmal* cosmos, world or universe, has been touched upon, commented on, or, stated, in a threadbare, prosaic, or, matter-of-fact, manner above, and will be dealt with, handled, or discussed in a more transcendent, acroamatic, or, abstract manner, a bit later. But before that one requires to restate what one already has said earlier, with regards to the site, location, or, place of residence, existence, presence, or, being, of the second, *dreamal* or *phantasmal,* cosmos, world, or universe.

The second, *dreamal* or *phantasmal,* cosmos, world or universe is defined as that *dreamal* or *phantasmal,* cosmos, world or universe which, willy-nilly, perforce, without a choice, whether desired or not, or, every which way, each and every day, is encountered, met, or, faced plus observed, seen, visualized, watched, felt, palpated, touched, tasted, heard, and smelt, that is to say, perceived and experienced, *fully, completely, all out,* or, *in all respects,* i.e., *to the fullest extent,* and, which, willy-nilly, perforce, without a choice, whether desired or not, or, every which way, each and every day, is participated in, or, taken part in, *fully, completely, all out,* or, *in all respects,* i.e., *to the fullest extent,* by human being's consciousness, each and every day, during its wakeful state.

This second, *dreamal* or *phantasmal,* cosmos, world or universe, exists, occurs, or obtains inside the unusual or uncommon, 3-D or three-dimensional form or version or, inside the unusual or uncommon, expanded, distended, dilated, or inflated form or version of the dimensionless *state* or *mode* of being, or, the unexpanded, undistended, undiluted, or uninflated *state* or *mode of being ,* or, the original *state* or *mode* of being, of the consciousness of God, Brahman, Daydreamer, Oneiricker, Author, Maker, Creator, Progenitor, Primogenitor, or Begetter the two *dreamal* or *phantasmal* cosmoses, worlds or universes which are the focus of one's attention or the subject of one's thought at the present moment, namely the dream sleep state cosmos, world, or, universe, on one hand, and the wakeful state cosmos, world, or, universe, on the other of human being's consciousness , both of which materialize without fail each 24 hours everywhere from nowhere during the entire life of human being's consciousness and which cease to exist, as for as human being's consciousness is concerned, only when the *physical, material,* or *concrete* looking and feeling body, whose presence human being's consciousness feels around itself during its wakeful state *only*, each and every day, finally says "good-bye" i.e., finally "dies" , "demises", "gives up the ghost", "gives up its ivy-like hold on the consciousness" , "kicks the bucket" , "relinquishes life" , "relinquishes consciousness" or "goes way of all flesh" and human being's consciousness coalesces back or merges back into the *infinite, boundless, limitless, omnipresent, omniscient, omnipotent, ubiquitous, universal, all-present,* or, *all-pervasive Ocean* of *Consciousness* or *Ocean* of *Mind* or, *Field* of *Consciousness* or *Field* of *Mind* , which is erroneously called, named, labeled, or tagged as Cosmic Space or Brahmandic Aakash by human being's consciousness during its wakeful state due to its nescience but which, in absolute reality , is the unusual or uncommon, 3-D or three-dimensional form or version or, the unusual or uncommon, expanded, distended, dilated, or inflated form or version of the usually, or, customarily, dimensionless form or version or, the usually or, customarily, unexpanded, undistended, undilated, or uninflated form or version of the *consciousness* of God or Brahman or, the consciousness of the Daydreamer, Oneiricker, Author, Maker, Creator, Progenitor, Primogenitor, or Begetter of the *physical, material,* or, *concrete* looking or feeling cosmos, world, or, universe of the wakeful state of human being's consciousness, on one hand, and the *physical, material,* or, *concrete* looking or feeling cosmos, world, or, universe of the dream sleep state of human being's consciousness, on the other, both of which , in absolute reality , are nothing but *dreamal* or *phantasmal* in nature, and created or made by this God, Brahman, Daydreamer, Oneiricker, Author, Maker, Creator, Progenitor, Primogenitor, or Begetter, i.e., by this *infinite, boundless, limitless, omnipresent, omniscient, omnipotent, ubiquitous, universal, all-present,* or, *all-pervasive Ocean* of *Consciousness* or *Ocean* of *Mind* or, *Field* of *Consciousness* or *Field* of *Mind* , which is erroneously called, named, labeled, or tagged as Cosmic Space or Brahmandic Aakash by human being's consciousness during its wakeful state due to its nescience.

The unusual or uncommon, 3-D or three-dimensional form or version or, the unusual or uncommon, expanded, distended, dilated, or inflated form or version of the dimensionless *state* or *mode* of being, or, the unexpanded, undistended, undiluted, or uninflated *state* or *mode of being* , or, the original *state* or *mode* of being, of the consciousness of God or Brahman, inside which this second, *dreamal* or *phantasmal,* cosmos, world or universe, exists, occurs, or obtains , is described in the realm of Adwait Vedant , in the following manner :-

This unusual or uncommon, 3-D or three-dimensional form or version or, this unusual or uncommon, expanded, distended, dilated, or inflated form or version of the dimensionless *state* or *mode* of being, or, the unexpanded, undistended, undiluted, or uninflated *state* or *mode of being* , or, the original *state* or *mode* of being, of the consciousness of God or Brahman, inside which this second, *dreamal* or *phantasmal,* cosmos, world or universe, exists, occurs, or obtains is an amazing, awesome, incredible or unbelievable, " *infinite, boundless, limitless, omnipresent, omniscient, omnipotent, ubiquitous, universal, all-present,* or, *all-pervasive Ocean* of *Consciousness* or *Ocean* of *Mind* or, *Field* of *Consciousness* or *Field* of *Mind* ".

This " *infinite, boundless, limitless, omnipresent, omniscient, omnipotent, ubiquitous, universal, all-present,* or, *all-pervasive Ocean* of *Consciousness* or *Ocean* of *Mind* or, *Field* of *Consciousness* or *Field* of *Mind* " inside which the second, *dreamal* or *phantasmal,* cosmos, world or universe, exists, occurs, or obtains, is widely, generally, or popularly named or called Cosmic Space or Brahmandic Aakash.

In other words, the nom de plume, nom de guerre, AKA, alias, allonym, anonym, ananym, assumed name, false name, fictitious name, nickname, pen name, pseudonym, or, last but not least, the pretend name of the Author, Creator, Maker, Progenitor, Primogenitor, Begetter, God, Brahman, Daydreamer, or, Oneiricker, of this second, *dreamal* or *phantasmal,* cosmos, world or universe, i.e., the *physical, material,* or *concrete* looking and feeling cosmos, world, or universe of the wakeful state of human being's consciousness, is "Cosmic Space" or "Brahmandic Aakash" , about whom, or, if one prefers, about which, the *Absolute Truth,* is that, it is an amazing, awesome, incredible or unbelievable " *infinite, boundless, limitless, omnipresent, omniscient, omnipotent, ubiquitous, universal, all-present,* or, *all-pervasive Ocean* of *Consciousness* or *Ocean* of *Mind* or, *Field* of *Consciousness* or *Field* of *Mind* " inside which the *physical, material, or concrete* looking and feeling cosmos, world, or universe, of the wakeful state of human being's consciousness, consisting of countless moons, planets, stars, black holes, galaxies and the like are floating, wafting or levitating plus whirling, twirling, and spiraling non-stop as its *day-dream, oneiric, dreamry, imagery* or *fantasy* only, and nothing else.

This second, *physical, material, or concrete* looking and feeling, *dreamal* or *phantasmal,* cosmos, world or universe of the wakeful state of human being's consciousness, consisting of countless moons, planets, stars, black holes, galaxies and the like, has been floating, wafting or levitating plus whirling, twirling, and spiraling non-stop inside this amazing, awesome, incredible or unbelievable *Absolute Truth* , namely, the " *infinite, boundless, limitless, omnipresent, omniscient, omnipotent, ubiquitous, universal, all-present,* or, *all-pervasive Ocean* of *Consciousness* or *Ocean* of *Mind* or, *Field* of *Consciousness* or *Field* of *Mind* " , nom de plume, nom de guerre, AKA, or, pretend name, Cosmic Space or Brahmandic Aakash, as its *day-dream, oneiric, dreamry, imagery* or *fantasy* only and nothing else and has been doing so from the beginning of cosmic time and will continue to do so till the end of cosmic time.

This unusual or uncommon *3-D* or *three-dimensional* form or version or, this unusual or uncommon *expanded, distended, dilated,* or *inflated* form or version of the dimensionless *state* or *mode of being* , or, the unexpanded, undistended, undiluted, or uninflated *state* or *mode of being,* or, the original *state* or *mode of being* , of the consciousness of God or Brahman, which is described as *infinite, boundless, limitless, omnipresent, omniscient, omnipotent, ubiquitous, universal, all-present,* or, *all-pervasive Ocean* of *Consciousness* or *Ocean* of *Mind* or, *Field* of *Consciousness* or *Field* of *Mind* , alias, aka, or, nom de guerre, Cosmic Space or

Brahmandic Aakash, or, which is widely, generally, or popularly named or called Cosmic Space or Brahmandic Aakash, is encountered, met or, faced, plus, observed, seen, visualized, or, watched by human being's consciousness during its wakeful state, with the aid of eyes of its *physical, material,* or, *concrete,* looking and feeling *body* of its wakeful state.

Human being's consciousness is, one hundred percent, o*bjectively* aware of the presence around itself, of its *physical, material,* or, *concrete,* looking and feeling, *dreamal* or *phantasmal* body, *only* during its wakeful state, each and every day.

However, during its dream sleep state, human being's consciousness, is one hundred percent *subjectively* unaware of the presence around itself, of its *physical, material, or concrete* looking and feeling, *dreamal* or *phantasmal* body, each and every day.

As a result, temporarily, during its dream sleep state *only,* each and every day, human being's consciousness exists, occurs, or, obtains as an amazing, awesome, incredible or unbelievable, bodiless consciousness, purely, simply, or only from its own , exclusive , private , or, personal, *subjective,* point of view, in the way , manner , style , demeanor , semblance , or, countenance of the *consciousness* of God or Brahman or, in the way , manner, , style , demeanor , semblance , or, countenance of the consciousness of the Daydreamer, Oneiricker, Author, Maker, Creator, Progenitor, Primogenitor, or Begetter of the *physical, material,* or, *concrete* looking or feeling, *dreamal* or *phantasmal* cosmos, world, or, universe of the wakeful state of human being's consciousness, on one hand, and of the *physical, material,* or, *concrete* looking or feeling *dreamal* or *phantasmal* cosmos, world, or, universe of the dream sleep state of human being's consciousness, on the other, that is to say, in the way , manner , style , demeanor semblance , or, countenance of Cosmic Space or Brahmandic Aakash or, in the way , manner , style , demeanor , semblance , or, countenance of the *infinite, boundless, limitless, omnipresent, omniscient, omnipotent, ubiquitous, universal, all-present,* or, *all-pervasive, Ocean* of *Consciousness* or *Ocean* of *Mind* or, *Field* of *Consciousness* or *Field* of *Mind* , whose alias, nom de plume, or, nom de guerre is Cosmic Space or Brahmandic Aakash, whom human being's consciousness *observes, sees, visualises,* or, *watches* during its wakeful state, each and every day, with the help of eyes of its *physical, material,* or, *concrete* looking or feeling *dreamal* or *phantasmal* body.

The *consciousness* of God or Brahman or, the consciousness of the Daydreamer, Oneiricker, Author, Maker, Creator, Progenitor, Primogenitor, or Begetter of the *physical, material,* or, *concrete* looking or feeling, *dreamal* or *phantasmal* cosmos, world, or, universe of the wakeful state of human being's consciousness, on one hand, and the *physical, material,* or, *concrete* looking or feeling *dreamal* or *phantasmal* cosmos, world, or, universe of the dream sleep state of human being's consciousness, on the other, is *eternally* bodiless.

That is to say, the *infinite, boundless, limitless, omnipresent, omniscient, omnipotent, ubiquitous, universal, all-present,* or, *all-pervasive Ocean* of *Consciousness* or *Ocean* of *Mind* or, *Field* of *Consciousness* or *Field* of *Mind* , which is widely, generally, or, popularly called, named, or labeled, Cosmic Space or Brahmandic Aakash, and which is observed, seen, visualised, or, watched, by human being's consciousness during its wakeful state, each and every day, with the aid of eyes of its *physical, material,* or, *concrete* looking and feeling *dreamal* or *phantasmal* body, is *eternally* bodiless.

Hence, during the limited period of its dream sleep state, each and every day, the *subjectively* bodiless *twin, clone, incarnation,* or *avatar* of human being's consciousness, *temporarily* becomes akin, analogous, or comparable to, or, carbon copy of, or, much the same as the consciousness of God or Brahman.

That is to say, during the limited period of its dream sleep state, each and every day, the *subjectively* bodiless *twin, clone, incarnation,* or *avatar* of human being's consciousness, *temporarily* becomes akin, analogous, or comparable to, or, carbon copy of, or, much the same as the consciousness of its Source or Spring.

In other words, during the limited period of its dream sleep state, each and every day, the *subjectively* bodiless *twin, clone, incarnation,* or *avatar* of human being's consciousness, *temporarily* becomes akin, analogous, or comparable to, or, carbon copy of, or, much the same as the consciousness of the Daydreamer, Oneiricker, Author, Maker, Creator, Progenitor, Primogenitor, or Begetter of the *physical, material,* or, *concrete* looking or feeling, *dreamal* or *phantasmal* cosmos, world, or, universe of the wakeful state of human being's consciousness, on one hand, and of the *physical, material,* or, *concrete* looking or feeling *dreamal* or *phantasmal* cosmos, world, or, universe of the dream sleep state of human being's consciousness, on the other.

And one must always keep in mind that God or Brahman, or, the Source or Spring of human being's consciousness, or, the Daydreamer, Oneiricker, Author, Maker, Creator, Progenitor, Primogenitor, or Begetter of the *physical, material,* or, *concrete* looking or feeling, *dreamal* or *phantasmal* cosmos, world, or, universe of the wakeful state of human being's consciousness, on one hand, and of the *physical, material,* or, *concrete* looking or feeling *dreamal* or *phantasmal* cosmos, world, or, universe of the dream sleep state of human being's consciousness, on the other, is none other than the *infinite, boundless, limitless, omnipresent, omniscient, omnipotent, ubiquitous, universal, all-present,* or, *all-pervasive Ocean* of *Consciousness* or *Ocean* of *Mind* or, *Field* of *Consciousness* or *Field* of *Mind* , who is widely, generally, or, popularly called, named, or labeled as Cosmic Space or Brahmandic Aakash, and who is *observed, seen, visualised,* or *watched by* human being's consciousness with the help or aid of eyes of its *physical, material,* or, *concrete* looking or feeling *dreamal* or *phantasmal* body during its wakeful state.

One more equally important point, one must also always keep in one's mind is the following :-

During the dream sleep state of human being's consciousness, latter's *consciousnessbally* embodied *twin, clone, incarnation* or *avatar*, or, if one prefers, latter's *objectively* embodied *twin, clone, incarnation* or, *avatar,* is noted, seen or spotted to wear, clad, don, or put on a consciousnessbal-substance-made, or a consciousnessbal-matter-made, *objective* body or, if one prefers, an objective-substance-made or, an objective-matter-made, *consciousnessbal* body . And this consciousnessbal-substance-made or consciousnessbal-matter-made, *objective* body or, if it is preferred, and this objective-substance-made or, objective-matter-made, *consciousnessbal* body of the *consciousnessbally* embodied or, if one prefers, *objectively* embodied *twin, clone, incarnation* or *avatar* of human being's consciousness, during latter's dream sleep state, possesses a set of five consciousnessbal-substance-made or, consciousnessbal-matter-made, *objective* sense organs or, if one prefers, a set of five objective-substance-made or, objective-matter-made *consciousnessbal* sense organs for the exclusive use of the *consciousnessbally* embodied *twin, clone, incarnation* or *avatar* human being's consciousness, or, if one prefers, for the exclusive use of the *objectively* embodied *twin, clone, incarnation* or, *avatar* of human being's consciousness, during latter's dream sleep state.

All the above described facts, truths, actualities, realities, or factualities, in connection with the *objectively* embodied *twin, clone, incarnation* or, *avatar* or, absolutely to the point, all the above described facts, truths, actualities, realities, or factualities, in connection with the *consciousnessbally* embodied *twin, clone, incarnation* or *avatar* of human being's consciousness, during latter's dream sleep state, is taken note of, or, spotted, by the *subjectively* bodiless *twin, clone, incarnation* or *avatar* of the same, very same, self-same, or, one and the same human being's consciousness, during latter's dream sleep state.

Then, all the above described facts, truths, or, realities in connection with the *objectively* embodied *twin, clone, incarnation* or, *avatar* or, absolutely to the point, then, all the above described facts, truths, or, realities in connection with the *consciousnessbally* embodied *twin, clone, incarnation* or *avatar* of human being's consciousness, which are noted or spotted by the *subjectively* bodiless *twin, clone, incarnation* or *avatar* of the same, very same, self-same, or, one and the same human being's consciousness, during latter's dream sleep state, are scrupulously, or assiduously plus steadfastly and permanently, retained in the memory of the *subjectively* bodiless *twin, clone, incarnation* or *avatar* of the same, very same, self-same, or, one and the same human being's consciousness for posterity, or, future use, or, for use at some future time or date, by the same, very same, self-same, or, one and the same *subjectively* bodiless *twin, clone, incarnation* or *avatar* of human being's consciousness of latter's dream sleep state.

Hence, the *subjectively* bodiless *twin, clone, incarnation* or *avatar* of human being's consciousness, which exists, occurs, or, obtains during latter's dream sleep state, remembers all the above described facts, truths, or, realities in connection with the *objectively* embodied *twin, clone, incarnation* or, *avatar* or, absolutely to the point, in connection with the *consciousnessbally* embodied *twin, clone, incarnation* or *avatar* of the same, selfsame, very same, or one and the same human being's consciousness, when it wakes up from sleep and becomes conscious, aware, or cognisant of its *objective* embodied-ness, that is to say, and becomes conscious, aware, or cognisant of its *physical, material,* or *concrete* looking and feeling embodied-ness, along with the consciousness, awareness, or cognisance of the objective cosmos, world, or universe of its wakeful state, that is to say, along with the consciousness, awareness, or cognisance of the *physical, material,* or *concrete* looking and feeling *cosmos, world,* or *universe* of its wakeful state.

What has been said above, can be put in a different way.

During the dream sleep state of human being's consciousness, each and every day, latter's *objectively* embodied *twin, clone, incarnation* or, *avatar* or, absolutely to the point, latter's *consciousnessbally* embodied *twin, clone, incarnation* or *avatar*, is noted, by the *subjectively* bodiless *twin, clone, incarnation* or *avatar* of the same, very same, self-same, or, one and the same human being's consciousness, to wear, clad, drape, don, deck, or, bedeck a consciousnessbal-substance-made or consciousnessbal-matter-made *body,* a consciousnessbal-substance-made or consciousnessbal-matter-made *body* which possesses five consciousnessbal-substance-made or, five consciousnessbal-matter-made *sense organs* for the exclusive use during the dream sleep state of human being's consciousness.

This *objectively* embodied *twin, clone, incarnation* or, *avatar* or, absolutely to the point, this *consciousnessbally* embodied *twin, clone, incarnation* or *avatar* of human being's consciousness, during latter's dream sleep state, willy-nilly, perforce, without a choice, whether desired or not, or, every which way, each and every day, encounters, meets, or, faces plus observes, sees, visualises, watches, feels, palpates, touches, tastes, hears, and smells, that is to say, perceives and experiences, *fully, completely, all out,* or, *in all respects,* i.e., *to the fullest extent,* and willy-nilly, perforce, without a choice, whether desired or not, or, every which way, each and every day, participates in or takes part in *fully, completely, all out,* or, *in all respects,* i.e., *to the fullest extent,* in all the *dreamal* or *phantasmal activities* of the *dreamal* or *phantasmal* cosmos, world, or universe of the *dream sleep state* of human being's consciousness, with the aid or assistance of its consciousnessbal-substance-made or consciousnessbal-matter-made *body,* on one hand, and with the aid or assistance of its body's five consciousnessbal-substance-made or consciousnessbal-matter-made *sense organs,* on the other.

The *objectively* embodied or, the *physical, material,* or, *concrete* looking and feeling *embodied* form, version, incarnation , or avatar of human being's consciousness, which *exists, obtains,* or, *occurs* during latter's wakeful state and which too willy-nilly, perforce, without a choice, whether desired or not, or, every which way, each and every day encounters, meets, or, faces plus observes, sees, visualises, watches, feels, palpates, touches, tastes, hears, and smells, that is to say, perceives and experiences, *fully, completely, all out,* or, *in all respects,* i.e., *to the fullest extent,* and, which too willy-nilly, perforce, without a choice, whether desired or not, or, every which way, each and every day participates in, or, takes part in, *fully, completely, all out,* or, *in all respects,* i.e., *to the fullest extent,* each and every day without fail, as well as participates in, or, takes part in, in all the *dreamal* or *phantasmal activities* of the *dreamal* or *phantasmal* cosmos, world, or universe during the wakeful state of human being's consciousness, with the assistance of its consciousnessbal-substance-made or consciousnessbal-matter-made *body* and this *body's* five consciousnessbal-substance-made or consciousnessbal-matter-made *sense organs,* during the wakeful state of human being's consciousness, each and every day, is alike, analogous, akin, or comparable to, or, is a carbon copy of, or, is much the same as the human being's consciousness who is *objectively* embodied *twin, clone, incarnation* or, *avatar* or, absolutely to the point, who is *consciousnessbally* embodied *twin, clone, incarnation* or *avatar* of human being's consciousness which exists, obtains, or occurs during latter's dream sleep state.

However, the consciousnessbal-substance-made or consciousnessbal-matter-made *body* whose presence around itself, human being's consciousness, willy-nilly, perforce, without a choice, whether desired or not, or, every which way, each and every day, encounters, or, faces, plus, observes, sees, visualises, watches, feels, palpates, touches, tastes, hears, and smells, that is to say, perceives and experiences, *fully, completely, all out,* or, *in all respects,* i.e., *to the fullest extent,* with the aid or assistance of the five consciousnessbal-substance-made or consciousnessbal-matter-made *sense organs* of its consciousnessbal-substance-made or consciousnessbal-matter-made *body,* and also, with the aid or assistance of the latter and its five consciousnessbal-substance-made or consciousnessbal-matter-made *sense organs,* it also willy-nilly, perforce, without a choice, whether desired or not, or, every which way, each and every day participates in, or, takes part in, *fully, completely, all out,* or, *in all respects,* i.e., *to the fullest extent,* each and every day without fail, in all the *dreamal* or *phantasmal activities* of the *dreamal* or *phantasmal* cosmos, world, or universe during its wakeful state, are nesciently or ignorantly thought or believed to be by human being's consciousness to be one hundred percent actual, factual, real , genuine, or, authentic, and is called or labeled by human being's conscious as one hundred percent *physical, material,* or *concrete,* just as human being's consciousness also nesciently or ignorantly thinks or believes that the *cosmos, world,* or, *universe* which it perceives and experiences during its wakeful state is also one hundred percent actual, factual, real , genuine, or, authentic, plus *physical, material,* or *concrete.*

Human being's consciousness is aware of the presence, around itself, of its *physical, material,* or, *concrete,* looking and feeling *body*, during its wakeful state only, and not during its dream sleep state. Isn't it astonishing? Does the existence of this amazing, astonishing, or, awesome, divaricate, or, "going separate ways" phenomenon, fact, wonder, spectacle, miracle, or, marvel, convey a hint to human being's consciousness with regards to the inherent, intrinsic, deep-down, absolutely true, or, through and through *nature* of the *physical, material,* or, *concrete* looking and feeling *cosmos, world,* or, *universe* which it willy-nilly, perforce, without a choice, whether desired or not, or, every which way, each and every day, encounters, meets, or, faces plus observes, sees, visualises, watches, feels, palpates, touches, tastes, hears, and smells, that is to say, perceives and experiences, *fully, completely, all out,* or, *in all respects,* i.e., *to the fullest extent,* and, in which, it willy-nilly, perforce, without a choice, whether desired or not, or, every which way, each and every day, participates in, or, takes part in, *fully, completely, all out,* or, *in all respects,* i.e., *to the fullest extent,* each and every day, during its wakeful state ?

In absolute contrast to its wakeful state during which human being's consciousness is aware of the presence around itself of its *physical, material,* or, *concrete,* looking and feeling *body,* that is to say, in absolute contrast to its wakeful state during which human being's consciousness is aware that it is in its *physical, material,* or, *concrete* looking and feeling, *embodied* form or version, or, in its *embodied* incarnation, or avatar, during its dream sleep state, human being's consciousness exists, occurs, or obtain in its bodiless form or version or, in its bodiless *incarnation* or *avatar.*

This bodiless form or version or, this bodiless *incarnation* or *avatar* of human being's consciousness, which exists, occurs, or, obtains during the dream sleep state of human being's consciousness, willy-nilly, perforce, without a choice, whether desired or not, or, every which way, each and every day, encounters, meets, or faces, plus, observes, sees, visualises, or, watches the *physical, material,* or, *concrete* looking and feeling cosmos, world, or, universe of its dream sleep state, willy-nilly, perforce, without a choice, whether desired or not, or, every which way, each and every day.

Additionally, this bodiless form or version or, this bodiless *incarnation* or *avatar* of human being's consciousness, which exists, occurs, or, obtains during latter's deep sleep state, has not an iota, jot, whit, or, scintilla of doubt with regards to the absolute *actuality, factuality, reality, authenticity,* or, *genuineness* of the *physical, material,* or, *concrete* looking and feeling cosmos, world, or, universe which it willy-nilly, perforce, without a choice, whether desired or not, or, every which way, each and every day, encounters, meets, or faces, plus, observes, sees, visualises, or, watches in its bodiless form or version or, in its bodiless *incarnation* or *avatar,* during its dream sleep state.

It is only on waking up from its sleep, slumber, torpor, sleep state or, slumberland that the human being's consciousness accepts, accedes, concedes, admits, acknowledges, recognises, or, own up to its judgmental folly, or, error, which it commits during its dream sleep state each and every day with regards to the true nature of the *physical, material, or concrete* looking and feeling *dreamal* or *phantasmal* cosmos, world, or, universe, which it regards to absolutely real, physical, material, or, concrete during its dream sleep state, and which it willy-nilly, perforce, without a choice, whether desired or not, or, every which way, each and every day, encounters, meets, or faces, plus, observes, sees, visualises, or, watches, willy-nilly, perforce, without a choice, whether desired or not, or, every which way, each and every day, in its bodiless form or version or, in its bodiless *incarnation* or *avatar* during its dream sleep state, each and every day, just as it also commits a judgmental folly, or, error, during its wakeful state with regards to the true nature of the *physical, material, or concrete* looking and feeling *dreamal* or *phantasmal* cosmos, world, or, universe, which it also regards to be absolutely real, physical, material, or, concrete during its wakeful state, willy-nilly, perforce, without a choice, whether desired or not, or, every which way, each and every day, encounters, meets, or faces, plus, observes, sees, visualises, watches, feels, palpates, touches, tastes, hears, and smells, that is to say, perceives and experiences, *fully, completely, all out,* or, *in all respects,* i.e., *to the fullest extent,* and, in which, it willy-nilly, perforce, without a choice, whether desired or not, or, every which way, each and every day, participates in, or, takes part in, *fully, completely, all out,* or, *in all respects,* i.e., *to the fullest extent,* each and every day but this time in its *physically, materially,* or *concretely* looking or feeling *embodied* form or version or, in its embodied *incarnation* or *avatar,* during its wakeful state, each and every day.

It is true that during its dream sleep state, human being's

consciousness, willy-nilly, perforce, without a choice, whether desired or not, or, every which way, each and every day, encounters, meets, or faces, plus, observes, sees, visualises, or, watches a *physical, material, or concrete* looking and feeling *dreamal* or *phantasmal* cosmos, world, or, universe while it is in its *subjectively* bodiless form or version or, while it is in its *subjectively* bodiless *incarnation* or *avatar* as compared to its wakeful state during which it willy-nilly, perforce, without a choice, whether desired or not, or, every which way, each and every day, encounters, meets, or faces, plus, observes, sees, visualises, or, watches a *physical, material, or concrete* looking and feeling *dreamal* or *phantasmal* cosmos, world, or, universe while it is in its *subjectively* embodied form or version or, while it is in its *subjectively* embodied *incarnation* or *avatar,* which it believes or think, albeit erroneously, that, it is in its *objectively* embodied form or version or, it is its *objectively* embodied *incarnation* or *avatar.* Here, human being's consciousness must recall that it makes the same, similar, or identical mistake, error, blunder, or gaffe during its dream sleep state too, again and again, each & every day, with regards to its *subjectively* embodied *twin, clone* , *incarnation* or *avatar* or, with regards to its *consciousnessbally* embodied *twin* , *clone* , *incarnation* or *avatar,* about which it too erroneously thinks or believes, while it is in its dream sleep state, that its *subjectively* embodied *twin* , *clone* , *incarnation* or *avatar* is in its *objectively* embodied form or version or, it is in its *objectively* embodied *incarnation* or *avatar.*

One more point.

As said above, human being's consciousness, during its dream sleep state, each and every day, exists, obtains, or, occurs in two *forms* or *versions* or, in two *incarnations* or *avatars* and not only one *form* or *version* or, not only one *incarnation* or, *avatar,* as is the case during its wakeful state when it exists, occurs, or, obtains in only one *form* or *version* or, in only one *incarnation* or, *avatar.*

Now the important question which human being's consciousness faces during its wakeful state is the following :-

"What is the significance of this amazing, astonishing, or, awesome, divaricate, or, "going separate ways" phenomenon, fact, wonder, spectacle, miracle, or, marvel?

That is to say, "What is the significance of the amazing, astonishing, or, awesome, divaricate, or, "going separate ways" phenomenon, fact, wonder, spectacle, miracle, or, marvel that during its dream sleep state, each and every day, human being's consciousness exists, obtains, or, occurs in its two *forms* or *versions* or, in its two *incarnations* or, *avatars* and not only one *form* or *version* or, only one *incarnation* or, *avatar,* as is the case during its wakeful state.

Let one explain.

During its wakeful state human being's consciousness exists, occurs, or, obtains in only one *form* or *version* or, in only one *incarnation* or, *avatar* i.e., during its wakeful state human being's consciousness exists, occurs, or, obtains in *subjectively* embodied *form* or *version* only or, in *subjectively* embodied *incarnation* or, *avatar* only.

Whereas in its dream sleep state, human being's consciousness exists, obtains, or, occurs in its *subjectively* bodiless *form* or *version* or, in its *subjectively* bodiless *incarnation* or, *avatar,* on one hand, and in its *subjectively* embodied *form* or *version* or, in its *subjectively* embodied *incarnation* or, *avatar,* on the other.

Therefore, the important question which human being's consciousness faces during its wakeful state is :-

"What is the earth-shattering significance of the above mentioned amazing, astonishing, or, awesome, divaricate, or, "going separate ways" phenomenon, fact, wonder, spectacle, miracle, or, marvel which human being's consciousness encounters, meets , or faces each and every day till one day its *physical, material,* or, *concrete* looking and feeling body, finally says "good-bye" i.e., finally "dies", "demises", "gives up the ghost" , "gives up its ivy-like hold on the consciousness" , "kicks the bucket" , "relinquishes life" , "relinquishes consciousness" or "goes way of all flesh" and human being's consciousness coalesces back or merges back into the *infinite, boundless, limitless, omnipresent, omniscient, omnipotent, ubiquitous, universal, all-present,* or, *all-pervasive Ocean* of *Consciousness* or *Ocean* of *Mind* or, *Field* of *Consciousness* or *Field* of *Mind* , which is erroneously called, named, labeled, or tagged as Cosmic Space or Brahmandic Aakash by human being's consciousness during its wakeful state due to its nescience but which, in absolute reality , is the unusual or uncommon, 3-D or three-dimensional form or version or, the unusual or uncommon, expanded, distended, dilated, or inflated form or version of the usually, or, customarily, dimensionless form or version or, the usually or, customarily, unexpanded, undistended, undilated, or uninflated form or version of the *consciousness* of God or Brahman or, the consciousness of the Daydreamer, Oneiricker, Author, Maker, Creator, Progenitor, Primogenitor, or Begetter of the *physical, material,* or, *concrete* looking or feeling cosmos, world, or, universe of the wakeful state of human being's consciousness, on one hand, and the *physical, material,* or, *concrete* looking or feeling cosmos, world, or, universe of the dream sleep state of human being's consciousness, on the other, both of which , in absolute reality , are nothing but *dreamal* or *phantasmal* in nature, and created or made by this God, Brahman, Daydreamer, Oneiricker, Author, Maker, Creator, Progenitor, Primogenitor, or Begetter, i.e., by this *infinite, boundless, limitless, omnipresent, omniscient, omnipotent, ubiquitous, universal, all-present,* or, *all-pervasive Ocean* of *Consciousness* or *Ocean* of *Mind* or, *Field* of *Consciousness* or *Field* of *Mind* , which is erroneously called, named, labeled, or tagged as Cosmic Space or Brahmandic Aakash by human being's consciousness during its wakeful state due to its nescience.

Let one explain "What is the earth-shattering significance of the above mentioned, amazing, astonishing, or, awesome, divaricate, or, "going separate ways" phenomenon, fact, wonder, spectacle, miracle, or, marvel?

The incredible *bodiless form, version, incarnation,* or, *avatar* of human being's consciousness, exists, occurs, or, obtains during latter's dream sleep state, each and every day, just as there exists, occurs, or, obtains during latter's wakeful state, its *physically, materially,* or, *concretely* looking and feeling *embodied form, version, incarnation,* or, *avatar,* each and every day, on one hand, and its *physically, materially,* or, *concretely* looking and feeling *embodied form, version, incarnation,* or, *avatar* of its dream sleep state, on the other.

To recap.

Human being's consciousness during its dream sleep state exists, occurs, or, obtains in two *forms, versions, incarnations,* or, *avatars,* namely, *bodiless form, version, incarnation,* or, *avatar,* on one hand, and *embodied form, version, incarnation,* or, *avatar,* on the other. During its wakeful state, human being's consciousness exists, occurs, or, obtains in its *physically, materially,* or, *concretely* looking and feeling *embodied form, version, incarnation,* or, *avatar* only.

Hence, there exists, or occurs, two *physically, materially,* or, *concretely* looking and feeling *embodied* forms, versions, incarnations, or, avatars of human being's consciousness, each and every day. One of them exists, occurs, or obtains during the dream sleep state of human being's consciousness and the other, exists, occurs, or obtains during the wakeful state of human being's consciousness.

The *physically, materially,* or, *concretely* looking and

feeling *embodied* form, version, incarnation, or, avatar of human being's consciousness, which exists, occurs, or obtains during the dream sleep state of human being's consciousness each and every day, willy-nilly, perforce, without a choice, whether desired or not, or, every which way, each and every day, observes, sees, visualises, watches, feels, palpates, touches, tastes, hears, and smells, that is to say, perceives and experiences, *fully, completely, all out,* or, *in all respects,* i.e., *to the fullest extent,* and, in it, it also willy-nilly, perforce, without a choice, whether desired or not, or, every which way, each and every day, participates, or, takes part *fully, completely, all out,* or, *in all respects,* i.e., *to the fullest extent* during the dream sleep state of human being's consciousness.

Now, the million-dollar question is "what is the role of the incredible *bodiless form, version, incarnation,* or, *avatar* of human being's consciousness which exists, occurs, or, obtains during the latter's dream sleep state *only*, each and every day, and not during its wakeful state. Furthermore, " what is the significance of the fact that this incredible *bodiless form, version, incarnation,* or, *avatar* of human being's consciousness which exists, occurs, or, obtains during latter's dream sleep state *only* , each and every day, and not during its wakeful state, willy-nilly, perforce, without a choice, whether desired or not, or, every which way, each and every day, encounters, meets, or, faces plus merely, purely, or, only, observes, sees, visualises or, watches with the aid of its consciousnessbal power of observing, seeing, visualising, or, watching, as it is bodiless i.e., as it does not possess a *physical, material, concrete body,* and therefore, does not possess *physical, material, concrete* eyes.

However, this incredible *bodiless form, version, incarnation,* or, *avatar* of human being's consciousness which exists, occurs, or, obtains during latter's dream sleep state, does not feel, palpate, touch, taste, hear, and smell, that is to say, does not perceive and experiences, *fully, completely, all out,* or, *in all respects,* i.e., *to the fullest extent,* and, it neither participates, or, takes part in any way or, in any which way.

The job of feeling, palpating, touching, tasting, smelling, and hearing, that is to say, perceiving and experiencing in each and every way, or, every which way or, to the fullest extent, and of taking part or, participating each and every way or every which way or to the fullest extent, in this *dreamal* or, *phantasmal* cosmos, world, or, universe, of the dream sleep state of human being's consciousness, is apportioned or, allotted or, given to the consciousnessbally *embodied, or,* to the *physically, materially, or, concretely,* looking and feeling *embodied* form, version, incarnation, or, avatar of human being's consciousness which exists, occurs, or obtains during the dream sleep state of human being's consciousness, by God, Brahman, Daydreamer, Oneiricker, Author, Maker, Creator, Progenitor, Primogenitor, or Begetter, i.e., by *infinite, boundless, limitless, omnipresent, omniscient, omnipotent, ubiquitous, universal, all-present,* or, *all-pervasive Ocean* of *Consciousness* or *Ocean* of *Mind* or, *Field* of *Consciousness* or *Field* of *Mind* , which is erroneously called, named, labeled, or tagged as Cosmic Space or Brahmandic Aakash by human being's consciousness during its wakeful state due to its nescience, and who is the Daydreamer, Oneiricker, Author, Maker, Creator, Progenitor, Primogenitor, or Begetter, of the the *dreamal* or, *phantasmal* cosmos, world, or, universe which exists, occurs, or, obtains during the dream sleep state of human being's consciousness.

On the other hand, the *physically, materially,* or, *concretely* looking and feeling *embodied* form, version, incarnation, or, avatar of human being's consciousness, which exists, occurs, or obtains during its wakeful state, each and every day, willy-nilly, perforce, without a choice, whether desired or not, or, every which way, each and every day, observes, sees, visualises, watches, feels, palpates, touches, tastes, hears, and, smells, that is to say, perceives and experiences, *fully, completely, all out,* or, *in all respects,* i.e., *to the fullest extent,* and, in it, it also willy-nilly, perforce, without a choice, whether desired or not, or, every which way, each and every day, participates, or, takes part *fully, completely, all out,* or, *in all respects,* i.e., *to the fullest extent* during its wakeful state.

Now, let's return to the dream sleep state of human being's consciousness and discuss the significance of its existence each and every day in the life of human being's consciousness.

In contrast to the second *dreamal* or *phantasmal* cosmos, world or universe of wakeful state of human being's consciousness, the first *dreamal* or *phantasmal* cosmos, world or universe, which exists, occurs, or obtains inside the unusual or uncommon, 3-D or three-dimensional form or version or, inside the unusual or uncommon, expanded, distended, dilated, or inflated form or version of the *subjectively* bodiless, *twin, clone, incarnation, or avatar* of human being's consciousness, during the latter's dream sleep state, contrary to the popular belief amongst human beings consciousnesses, is created, or, brought into being, inside the unusual or uncommon, expanded, distended, dilated, or inflated form or version of the *subjectively* bodiless, *twin, clone, incarnation, or avatar* of human being's consciousness, during latter's dream sleep state, by no other consciousness than the same, very same, selfsame, or, one and the same consciousness of

God or *Brahman,* with the sole purpose of apprising, advising, enlightening, or, informing, human being's consciousness during its wakeful state when the latter is in its *physical, material, or, concrete* looking and feeling, *embodied* form or version, that the physical, material, substantial, or concrete looking and feeling cosmos, world or universe, which human being's consciousness, in its *physical, material, or, concrete* looking and feeling, *embodied* form or version, during its wakeful state, observes, sees, visualizes, watches, feels, palpates, touches, tastes, hears, and smells, that is to say, perceives and experiences, *fully, completely, all out,* or, *in all respects,* i.e., *to the fullest extent,* and in which, it participates, or, takes part *fully, completely, all out,* or, *in all respects,* i.e., *to the fullest extent,* is also of *dreamal* or *phantasmal* nature and nothing else.

In other words, the first *dreamal* or *phantasmal* cosmos, world or universe, which, as said above, exists, occurs, or obtains inside the unusual or uncommon, 3-D or three-dimensional form or version or, inside the unusual or uncommon, expanded, distended, dilated, or inflated form or version of the *subjectively* bodiless, *twin, clone, incarnation, or avatar* of human being's consciousness, during the latter's dream sleep state, contrary to the popular belief amongst the human beings consciousnesses, is created, or, brought into being, inside the unusual or uncommon, expanded, distended, dilated, or inflated form or version of the *subjectively* bodiless, *twin, clone, incarnation, or avatar* of human being's consciousness, during the latter's dream sleep state, by no other consciousness than the same, very same, selfsame, or, one and the same *omnipresent, all-present, all-pervading, universal, ubiquitous,* or, *infinite* Field of Consciousness called Cosmic Space or Brahmandic Aakash, with the sole purpose of apprising, advising, enlightening, or, informing, human being's consciousness during its wakeful state when the latter is in its *physical, material, or, concrete* looking and feeling, *embodied* form or version, that the physical, material, substantial, or concrete looking and feeling cosmos, world or universe, which the human being's consciousness, in its *physical, material, or, concrete* looking and feeling, *embodied* form or version, during its wakeful state, observes, sees, visualizes, watches, feels, palpates, touches, tastes, hears, and smells, that is to say, perceives and experiences, *fully, completely, all out,* or, *in all respects,* i.e., *to the fullest extent,* and, in which, it participates, or, takes part *fully, completely, all out,* or, *in all respects,* i.e., *to the fullest extent,* is also of *dreamal* or *phantasmal* nature and nothing else.

The above described, conscious, deliberate, intentional, proactive, or thoughtful action, deed, or gesture on the part of consciousness of God, Brahman viz., on the part of the *omnipresent, all-present, all-pervading, spreading through or into everything, universal, ubiquitous,* or, *infinite* Field of Consciousness called Cosmic Space or Brahmandic Aakas, namely, of apprising, advising, enlightening, or informing human being's consciousness during its wakeful state, when the latter is in its *physical, material, or, concrete* looking and feeling, *embodied* form or version, that the *physical, material, or, concrete* looking and feeling cosmos, world or universe, which human being's consciousness in its *physical, material, or, concrete* looking and feeling, *embodied* form or version, during its wakeful state, observes, sees, visualizes or watches, plus, feels or touches, and hears, smells, and tastes, that is to say, perceives and experiences as well as in which it participates or, takes part, is also of the *dreamal* or *phantasmal* nature and nothing else, is undertaken, engaged in, or taken up, by the consciousness of God, Brahman viz., is undertaken, engaged in, or taken up, by the *omnipresent, all-present, all-pervading, spreading through or into everything, universal, ubiquitous,* or, *infinite* Field of Consciousness called Cosmic Space or Brahmandic Aakash because amongst all the *conscious, aware* or *sentient,* embodied beings that have been *dreamed up, invented, actualized,* or, *brought into being* by God or Brahman inside its consciousness through the agency, means, or, instrumentality of daydreaming or oneiricking on its part viz., which have been *dreamed up, invented, actualized,* or, *brought into being* by the *ubiquitous* and *infinite* field of consciousness called Cosmic Space or Brahmandic Aakash through the agency, means, or, instrumentality of daydreaming or oneiricking on its part, a human being and his human brain has been accorded, awarded, bestowed, conferred, or granted by it, the highest amount of consciousness, awareness or sentience and therefore, highest amount of *intelligence, intellectual capacity, insight, intuition, acumen, brainpower, brilliance, cleverness, comprehension, discernment, giftedness, judgement, penetration, power of reasoning, perceptiveness, quickness of mind, sharpness, smartness,* or *talent* .

To repeat.

The above described, conscious, deliberate, intentional, purposive, proactive, or thoughtful action, deed, or gesture on the part of consciousness of God, Brahman viz., on the part of the *ubiquitous and infinite* field of consciousness called Cosmic Space or Brahmandic Aakas, namely, of apprising, advising, enlightening, or informing human being's consciousness during its wakeful state, when the latter is in its *physical, material, or, concrete* looking and feeling, *embodied* form or version, that the *physical, material, or, concrete* looking and feeling cosmos, world or universe, which human being's consciousness in its *physical, material, or, concrete* looking and feeling, *embodied* form or version, during its wakeful state, observes, sees, visualises or watches, plus, feels or touches, and hears, smells, and tastes, that is to say, perceives and experiences as well as in which it participates or, takes part, is also of *dreamal* or *phantasmal* nature and nothing else, is undertaken, engaged in, or taken up, by the consciousness of God, Brahman viz., is undertaken, engaged in, or taken up, by the (*ubiquitous* and *infinite) omnipresent, all-present, all-pervading, spreading through or into everything, universal, ubiquitous,* or, *infinite* Field of Consciousness called Cosmic Space or Brahmandic Aakash because amongst all the *conscious, aware* or *sentient* embodied beings which have been *dreamed up, invented, actualised,* or, *brought into being* by God or Brahman inside its consciousness through the agency, means, or, instrumentality of daydreaming or oneiricking on its part viz., which have been *dreamed up, invented, actualised,* or, *brought into being* by the *ubiquitous* and *infinite omnipresent, all-present, all-pervading, spreading through or into everything, universal, ubiquitous,* or, *infinite* Field of Consciousness called Cosmic Space or Brahmandic Aakash through the agency, means, or, instrumentality of daydreaming or oneiricking on its part, a human being and his human brain has been accorded, awarded, bestowed, conferred, or granted by it, the highest amount of consciousness, awareness or sentience and therefore, highest amount of *intelligence, intellectual capacity, insight, intuition, acumen, brainpower, brilliance, cleverness, comprehension, discernment, giftedness, judgement, penetration, power of reasoning, perceptiveness, quickness of mind, sharpness, smartness,* or *talent* .

As a consequence of possessing the highest amount of consciousness, awareness, or sentience in its brain, and therefore, the highest amount of *intelligence, intellectual capacity, insight, intuition, acumen, brainpower, cleverness, brilliance, comprehension, discernment, giftedness, judgement, penetration, power of reasoning, perceptiveness, quickness of mind, sharpness, smartness,* or *talent,* human being's consciousness is the only one amongst all the *conscious, aware* or *sentient* scions, descendants, heirs, offsprings, or, progenies of the consciousness of God or Brahman viz., amongst all the *conscious, aware* or *sentient* scions, descendants, heirs, offsprings, or, progenies of the (*ubiquitous* and *infinite) omnipresent, all-present, all-pervading, spreading through or into everything, universal, ubiquitous,* or, *infinite* Field of Consciousness called Cosmic Space or Brahmandic Aakash, which is bedevilled, bothered, dogged, discomfited, harassed, harried, needled, nettled, pestered, plagued, riled, rankled, tormented, troubled, vexed, frustrated, or, miffed by such extremely difficult to answer questions as :-

"Who or what is its actual, real, or, true source or spring

i.e., who or what is the *actual, real, or, true source or spring* of human being's consciousness*?"*

"What for or why has it *come into being?"*

"How has it come into being?"

"Why did the physical, material, substantial, objective or *concrete cosmos, which it perceives, experiences,and in which it takes part during its wakeful state, was actualized or, brought into being and, who or what actualized it or, brought it into being?" plus*

"What is the true nature of the physical, material, substantial, objective or *concrete cosmos which it perceives, experiences, and in which it takes part during its wakeful state*"?

The above questions and many more of their ilk, type or, genre, bedevil, torment, or, harry human being's consciousness every day during its wakeful state, when it is in its *physical, material, or, concrete* looking and feeling *embodied* form or version, throughout its limited, finite, or, time-bound presence or existence in the physical, material, or concrete looking and feeling cosmos, world, or, universe.

In order to explain, enlighten, or expound, or, in order

to shed light on or make see daylight, plus, in order to assist human being's consciousnesses vis-a-vis the answers to all its above extremely difficult questions which it constantly poses to itself during its wakeful state, when it is in its *physical, material,* or, *concrete* looking and feeling *embodied* form or version, the consciousness of God or Brahman viz., the *ubiquitous* and *infinite* Field of Consciousness called Cosmic Space or Brahmandic Aakash, provides to each and every human being's consciousness, and, continues to provide to each and every human being's consciousness, during its absolutely necessary, mandatory, compulsory, vital, or, indispensable activity of sleep, each and every day, the *experience* of the mind-boggling or mind-blowing dream sleep state and the dream sleep state's, amazing, dazzling, or, stunning, *physical, material, or, concrete* looking and feeling *dreams* or the *physical, material, or, concrete* looking and feeling *dreamal* or *phantasmal cosmoses, worlds,* or, *universes,* each and every day, which each and every human being's consciousness during its dream sleep state, each and every day, observes, sees, visualises, watches, feels, palpates, touches, tastes, hears, and, smells, that is to say, perceives and experiences, *fully, completely, all out,* or, *in all respects,* i.e., *to the fullest extent,* and, in which, it participates, or, takes part *fully, completely, all out,* or, *in all respects,* i.e., *to the fullest extent,* with the aid or assistance or, through the instrumentality or agency of its two twins, clones, incarnations, or, avatars, namely, the *subjectively* bodiless twin, clone, incarnation, or, avatar, on one hand, and the *objectively,* or, absolutely to the point, *consciousnessbally* embodied twin, clone, incarnation, or, avatar, on the other, while each and every human being's consciousness during its mind-boggling or mind-blowing dream sleep state remains one hundred percent unaware, unconscious, unsuspecting, uninformed, unenlightened, unmindful, oblivious, in the dark, or, incognisant, i.e., possesses no knowledge or inkling whatsoever, of the existence or presence of its physical body, whose existence or presence, it is aware or conscious, one hundred percent, plus, all the while, around itself, during its wakeful states.

Here one will like to remind oneself anew or underscore that the *subjectively* bodiless, *twin, clone, incarnation, or avatar* of human being's consciousness during latter's dream sleep state, observes, sees, visualises or watches, plus experiences the amazing, dazzling, or, stunning, *physical, material, or, concrete* looking and feeling *dreams* or the *physical, material, or, concrete* looking and feeling *dreamal* or *phantasmal cosmoses, worlds,* or, *universes,* each and every day, during the dream sleep state of human being's consciousness, while it itself *subjectively* remains one hundred percent bodiless or without a body.

At this juncture it is also absolutely essential for one to remind oneself anew or underscore that the consciousness of God or Brahman i.e., the (ubiquitous and infinite) *omnipresent, all-present, all-pervading, spreading through or into everything, universal, ubiquitous,* or, *infinite* Field of Consciousness aka Cosmic Space or Brahmandic Aakash is eternally bodiless or is eternally without a body.

Hence, while each and every *subjectively* bodiless, *twin, clone, incarnation, or avatar* of human being's consciousness during latter's dream sleep state, observes, sees, visualises or watches, plus experiences the amazing, dazzling, or, stunning, *physical, material, or, concrete* looking and feeling *dreams* or the *physical, material, or, concrete* looking and feeling *dreamal* or *phantasmal cosmoses, worlds,* or, *universes,* each and every day, during the dream sleep state of human being's consciousness, it temporarily becomes like the consciousness of God or Brahman, i.e., it temporarily becomes like the *ubiquitous* and *infinite* field of consciousness called Cosmic Space or Brahmandic Aakash, with regards to its experience of bodiless-ness or non-embodied-ness, albeit or notwithstanding, *subjectively* only and not *actually* or *factually*, on one hand, and with regards to its becoming 3-D or three-dimensional in configuration or silhouette i.e., with regards to its becoming expanded, distended, dilated, or, inflated in configuration or silhouette plus with regards to the presence or existence of an amazing, dazzling, or, stunning, *physical, material, or, concrete* looking and feeling *dream* or the *physical, material, or, concrete* looking and feeling *dreamal* or *phantasmal cosmos, world,* or, *universe* inside its 3-D or three-dimensional form or version or inside its expanded, distended, dilated, or, inflated form or version of its consciousness, on the other.

To repeat.

Hence, while each and every *subjectively* bodiless, *twin, clone, incarnation, or avatar* of human being's consciousness during latter's dream sleep state, observes, sees, visualises or watches, plus experiences the amazing, dazzling, or, stunning, *physical, material, or, concrete* looking and feeling *dreams* or the *physical, material, or, concrete* looking and feeling *dreamal* or *phantasmal cosmoses, worlds,* or, *universes,* each and every day, during the dream sleep state of human being's consciousness, it temporarily becomes like the consciousness of God or Brahman, i.e., it temporarily becomes like the (*ubiquitous* and *infinite) omnipresent, all-present, all-pervading, spreading through or into everything, universal, ubiquitous,* or, *infinite* Field of Consciousness called Cosmic Space or Brahmandic Aakash, with regards to its experience of bodiless-ness or non-embodied-ness, albeit or notwithstanding, *subjectively* only and not *actually* or *factually*, on one hand, and with regards to its becoming 3-D or three-dimensional in configuration or silhouette i.e., with regards to its becoming expanded, distended, dilated, or, inflated in configuration or silhouette plus with regards to the presence or existence of an amazing, dazzling, or, stunning, *physical, material, or, concrete* looking and feeling *dream* or the *physical, material, or, concrete* looking and feeling *dreamal* or *phantasmal cosmos, world,* or, *universe* inside its 3-D or three-dimensional form or version or inside its expanded, distended, dilated, or, inflated form or version of its consciousness, on the other.

During the absolutely necessary, vital, or, indispensable activity of sleep, on the part of human being's consciousness, each and every day, the experience faced by the latter's *subjectively* bodiless *twin, clone, incarnation, or avatar,* of the mind-boggling phenomenon called dream sleep state and dream sleep state's, amazing, dazzling, or, stunning, *physical, material, or, concrete* looking and feeling *dreams* or, the *physical, material, or, concrete* looking and feeling *dreamal* or *phantasmal cosmoses, worlds,* or, *universes,* each and every day, during the dream sleep state of human being's consciousness, has no other raison d'être, reason, rationale, role, justification, or function, or, has no other basis or purpose than to explain, enlighten, expound, shed light on, or, to make see daylight, plus, to assist or aid human being's consciousnesses vis-a-vis the answers to all its earlier listed, extremely difficult questions which it constantly poses to itself during its wakeful state, when in its *physical, material,* or, *concrete* looking and feeling *embodied* form or version

Therefore, it is up to human being's consciousnesses to derive as much benefit as it wants from the daily *experience* of its mind-boggling or mind-blowing dream sleep state and dream sleep state's amazing, dazzling, or, stunning, *physical, material, or, concrete* looking and feeling *dreams* or, the *physical, material, or, concrete* looking and feeling *dreamal* or *phantasmal cosmoses, worlds,* or, *universes,* in order to find answers to all the above listed questions which it poses to itself on the daily basis during its wakeful state.

To sum up.

On one hand, through the aid or assistance of its *subjectively* bodiless, *twin, clone, incarnation, or avatar* of its dream sleep state, human being's consciousness, during its dream sleep state, perforce, willy-nilly, without a choice, whether desired or not, or, every which way, is made or obliged, purely to observe, see, visualise, or watch a "*phenomenon*" or "*magic-show*" called "*dream world", "fantasy world"*, *"dreamal world"* or *"fantasmal world"* every night, during its dream sleep state.

On the other hand, through the aid or assistance of its consciousnessbally *embodied* or, if one prefers, objectively *embodied* twin, clone, incarnation or avatar of its dream sleep state, human being's consciousness, during its dream sleep state, perforce, willy-nilly, without a choice, whether desired or not, or, every which way, is made or obliged, not purely to observe, see, visualise, or watch, but is also, perforce, willy-nilly, without a choice, whether desired or not, or, every which way, made or obliged to touch, taste, smell and hear, i.e., perceive and experience *fully, completely, all out,* or *in all respects,* the very same, one and the same or self-same *phenomenon* or *magic-show* called *"dream world", "fantasy world", "dreamal world"* or *"fantasmal world"* every night, during its dream sleep state through the *aid or assistance* of all the *five,* objective-looking but *consciousnessbal-substance-made sense organs,* of its objective-looking but *consciousnessbal-substance-made body.*

Not only that, through the aid or assistance of its consciousnessbally *embodied* twin, clone, incarnation or avatar or, if one prefers, objectively *embodied* twin, clone, incarnation or avatar of its dream sleep state, human being's consciousness, during its dream sleep state, perforce, willy-nilly, without a choice, whether desired or not, or, every which way, is also made or obliged to take part or participate *fully, completely, every inch,* or *all out,* in the very same, one and the same, or self-same *phenomenon* or *magic-show* called *"dream world", "fantasy world", "dreamal world"* or, *"fantasmal world"* every night, during its dream sleep state, through the *aid or assistance* of all the very same, one and the same, or self-same, *five* objective-looking but *consciousnessbal-substance-made,* sense organs of its very same, one and the same, or self-same objective-looking but *consciousnessbal-substance-made* body.

To recap.

The *subjectively* bodiless, *twin, clone, incarnation, or avatar* of human being's consciousness during the latter's dream sleep state, perforce, willy-nilly, without a choice, whether desired or not, or, every which way, is made or obliged to purely observe, see, visualise, and watch, a *phenomenon* or *magic-show* called *"dream world", "fantasy world", "dreamal world",* or, *"fantasmal world"* every night, during its dream sleep state, which it observes, sees, visualises, or watches with the aid or assistance of its *consciousnessbal* kind, type, or, variety of *observing, seeing, visualising,* or *watching* skill, talent, adroitness, competence, ability, facility or capacity and not with any kind, type, or, variety of physical, material, substantial, objective, or concrete *senses* or *sense organs* as it does during its wakeful state.

This *phenomenon* or the *magic-show* called "*dream world", "fantasy world", "dreamal world",* or, *"fantasmal world";* (which human being's consciousness, perforce, willy-nilly, without a choice, whether desired or not, or, every which way, is made or obliged to purely observe, see, visualise, or watch, during its dream sleep state through the *agency, medium,* or *vehicle* of its *subjectively* bodiless *twin, clone, incarnation* or *avatar* during its dream sleep state); is correctly, justly, or, rightly, labeled, branded or categorised as being nothing but a mere "dream", "dreamry", "imagery", or "fantasy" and therefore, as being nothing but only "partially real", "to a certain degree real " or, "real only when it obtains, occurs, or exists but not otherwise" i.e., as being nothing but only ephemeral, evanescent, episodic, fleeting, passing, short-lived, transient, temporary or impermanent, by human being's consciousness, not while it was observing, seeing, visualising, or watching it during its dream sleep state, but later on, following due deliberation and diligence on its part during its wakeful state when it once again becomes aware of its physical, material, substantial, or, concrete, *embodied-ness* on one hand, and of the physical, material, substantial, or, concrete *cosmos, world,* or *universe,* on the other.

To repeat.

Human being's consciousness, through the *agency, medium,* or *vehicle* of its *subjectively* bodiless *twin, clone, incarnation* or *avatar* during its dream sleep state, is made or obliged to purely observe, see, visualise, or, watch, perforce, willy-nilly, without a choice, whether desired or not, or, every which way, the *phenomenon* or the *magic-show* of its dream sleep state. It correctly, justly, or, rightly labels, brands, or categorises this *phenomenon* or the *magic-show* of its dream sleep state as being a mere "dream", "dreamry", "imagery", or, "phantasy" or, as being a mere "*dream world", "fantasy world*", *"dreamal world",* or, *"fantasmal world",* and therefore, only "partially real", "to a certain degree real" or, "real only when it obtains, occurs, or exists but not otherwise" or, "mithya" i.e., ephemeral, evanescent, episodic, fleeting, passing, short-lived, transient, temporary or impermanent.

However, human being's consciousness must, dwell upon, recognise, recall, recollect, remember, or, summon from its deepest recesses, the fact that, it is able to, or, becomes adjusted, disposed, inclined, minded, prepped, primed, prepared, psyched-up, rigged, set, willing, or wired to correctly, justly, or, rightly, label, brand, or categorise the *phenomenon* or the *magic-show* of its dream sleep state as being a mere "dream", "dreamry", "imagery", or, "phantasy" or, as being a mere "*dream world", "fantasy world", "dreamal world",* or, *"fantasmal world",* and therefore, only "partially real", "to a certain degree real" or, "real only when it obtains, occurs, or exists but not otherwise" or, "mithya" i.e., ephemeral, evanescent, episodic, fleeting, passing, short-lived, transient, temporary or impermanent, not while it remains occupied, engaged, engrossed, or head over heels in observing, seeing, visualising, or watching it through the *agency, medium,* or *vehicle* of its *subjectively* bodiless *twin, clone, incarnation* or *avatar,* but only later on, following the due deliberation and diligence on its part during its wakeful state or, following the due deliberation and diligence on its part, on waking up from its "dream", or, following the due deliberation and diligence on its part, when it once again becomes aware of its physical, material, substantial, or, concrete, *embodied-ness* on one hand and aware of the physical, material, substantial, or, concrete *cosmos, world,* or *universe* of its wakeful state, on the other. While it was occupied, engaged, engrossed, or head over heels in observing, seeing, visualising, or watching this mere "dream", "dreamry", "imagery", or, "phantasy" or, this mere *"dream world", "fantasy world", "dreamal world",* or, *"fantasmal world"* of its dream sleep state, through the *agency, medium,* or *vehicle* of its *subjectively* bodiless *twin, clone, incarnation* or *avatar* of its dream sleep state, it regarded this mere "dream", "dreamry", "imagery", or, "phantasy" or, this mere *"dream world", "fantasy world", "dreamal world",* or, *"fantasmal world"* of its dream sleep state as being *one hundred percent* genuine, authentic, real or, "satya", or, as being *one hundred percent* actual, factual, material, substantial, objective, solid, concrete, feelable, palpable, touchable, tastable, smellable and hearable. In that state of its mind, it was not in the least aware of the fact that this *cosmos, world,* or *universe* was a mere *"dream world", "fantasy world", "dreamal world",* or, *"fantasmal world",* and therefore, only "partially real", "to a certain degree real" or, "real only when it obtains, occurs, or exists but not otherwise" or, "mithya" i.e., ephemeral, evanescent, episodic, fleeting, passing, short-lived, transient, temporary or impermanent.

Exactly similar is the case, situation, or position, or, exactly similar is the truth, actuality, or, reality with regards to or vis-a-vis the "dream", "dreamry", "imagery", or, "phantasy" or, the *"dream world", "fantasy world", "dreamal world",* or, *"fantasmal world",* which human being's consciousness, during its wakeful state; through the *agency, medium,* or *vehicle* of its *physically, materially, substantially,* or *concretely* embodied, variant, form, version, or rendition; perforce, willy-nilly, without a choice, whether desired or not, or, every which way, is made or obliged to observe, see, visualise, watch, feel, touch, taste, hear and smell, as well as take part or participate in the *phenomenon* or the *magic-show* of its wakeful state which is illiterately, nesciently, ignorantly, or wrongly called physical, material, substantial, or concrete *cosmos, world,* or *universe* by human being's consciousness during its wakeful state, due to the lack of right and proper deliberation and diligence on its part.

The view or the opinion of some people that only physical, material, substantial, or concrete *phenomenon* of the universe can be known by human being's consciousness and knowledge of the ultimate cause of this physical, material, substantial, or concrete *phenomenon* of the universe on one hand, and the knowledge of the ultimate nature of cosmic space plus the knowledge of the ultimate nature of human being's consciousness, on the other, is unknowable or impossible by human being's consciousness, is an absolute hokum, disinformation, guile, or, mendacity plus it is a sign of closed-mindedness.

The "dream", "dreamry", "imagery", or, "phantasy" or, the *"dream world", "fantasy world"*, *"dreamal world",* or, *"fantasmal world";* which human being's consciousness, during its wakeful state; through the *agency, medium,* or *vehicle* of its *physically, materially, substantially,* or *concretely,* embodied, variant, form, version, or rendition; perforce, willy-nilly, without a choice, whether desired or not, or, every which way; is made or obliged to observe, see, visualise, watch, feel, touch, taste, hear and smell, as well as take part or participate in its wakeful state, also looks, appears, or seems as being *one hundred percent* genuine, authentic, real or, "satya", or, as being *one hundred percent* actual, factual, material, substantial, objective, solid, concrete, feelable, palpable, touchable, tasteable, smellable and hearable, while human being's consciousness; in its *physically, materially, substantially,* or *concretely* embodied, variant, form, version or rendition; is completely occupied, engaged, engrossed, or head over heels, both physically and mentally, in observing, seeing, visualising, watching, feeling, touching, tasting, hearing, smelling, and participating or taking part in it, without the application of due deliberation and due diligence on its part, with regards to its ultimate or quintessential nature. However, following the application of due deliberation and due diligence on the part of human being's consciousness with regards to the ultimate or quintessential nature of the *physical universe* of mankind's wakeful state, human being's consciousness during its wakeful state, is bound to discover or find out one day that, the physical, material, substantial, or concrete cosmos, world, or universe, which it, perforce, willy-nilly, without a choice, whether desired or not, or, every which way, is made or obliged to observe, see, visualise, watch, feel, touch, taste, hear and smell, as well as take part or participate in, during its wakeful state; while it is in its embodied, variant, form, version or rendition; is also a mere "dream", "dreamry", "imagery", or, "phantasy" or, a mere *"dream world", "fantasy world"*, *"dreamal world",* or, *"fantasmal world",* which is at par with or coequal with the "dream", "dreamry", "imagery", or, "phantasy" or, the *"dream world", "fantasy world"*, *"dreamal world",* or, *"fantasmal world"* which it observes, sees, visualises or watches, plus experiences every night during its dream sleep state, in its bodiless variant, form, version or rendition.

What has been said above can be put in another way.

The *phenomenon* or the *magic-show* of dream sleep state of human being's consciousness, which the latter i.e., human being's consciousness is made or obliged to purely observe, see, visualise, or, watch, perforce, willy-nilly, without a choice, whether desired or not, or, every which way, during its dream sleep state through the *agency, medium,* or *vehicle* of its *subjectively* bodiless *twin, clone, incarnation* or *avatar* of its dream sleep state, is correctly, justly, or, rightly labeled, branded, or categorised as being a mere "dream", "dreamry", "imagery", or, "phantasy" or, as being a mere "*dream world", "fantasy world"*, *"dreamal world",* or, *"fantasmal world",* and therefore, only "partially real", "to a certain degree real" or, "real only when it obtains, occurs, or exists but not otherwise" i.e., ephemeral, evanescent, episodic, fleeting, passing, short-lived, transient, temporary or impermanent, by human being's consciousness, not while it is occupied, engaged, engrossed, or head over heels in observing, seeing, visualising, or watching it, but only later on, following the due deliberation and diligence on its part during its wakeful state or when it once again becomes aware of its physical, material, substantial, or, concrete, *embodied-ness* on one hand and of the physical, material, substantial, or, concrete *cosmos, world*, or *universe,* on the other.

That is to say, the *phenomenon* or the *magic-show* called *"dream world", "fantasy world"*, *"dreamal world",* or, *"fantasmal world",* which human being's consciousness observes, sees, visualises, or watches during its dream sleep state, it correctly, justly, or rightly, labels, brands, or categorises as being a mere "dream", "dreamry", imagery", or "phantasy" or, as being a mere *"dream world", "phantasy world", "dreamal world"* or *"phantasmal world".*

However, human being's consciousness must always recall, recollect, recognise, remember, dwell upon, or, summon from its deepest recesses the fact that, the *phenomenon* or the *magic-show* called *"dream world", "fantasy world"*, *"dreamal world",* or, *"fantasmal world",* which it observes, sees, visualises, or watches during its dream sleep state, it correctly, justly, or, rightly labels, brands, or categorises as being a mere "dream", "dreamry", imagery", or "phantasy", not while it was observing, seeing, visualising, or, watching it, but later on, following the due deliberation or diligence on its part during its wakeful state or when it once again becomes aware of its *embodied-ness* of physical, material, substantial, or, concrete kind, variety, or type, on one hand and aware of the physical, material, substantial, or, concrete cosmos, *world,* or *universe,* on the other.

At this juncture it will be extremely useful to point out to oneself, that, the embodied-ness of human being's consciousness is of two kinds, types, or varieties, namely, of the physical, material, substantial, or, concrete kind, type, or variety, on one hand, and of the consciousnessbal kind, type, or variety, on the other.

Both these kinds, types, or, varieties of embodied-ness, look, seem or, appear objective in equal measure to two different forms, versions, or, variants, of the same, very same, selfsame, or, one and the same human consciousness, inside the confines or the boundaries of their respective ambiences, surroundings, milieus, backdrops, settings, locales, turfs, arenas, spheres, or domains.

Let one explain.

Human being's consciousness possesses the *embodied-*

ness of the physical, material, substantial, or, concrete kind during its wakeful state.

In stark contrast, the objectively *embodied* twin, clone, incarnation or avatar of human bcing's consciousncss during the latter's dream sleep state possesses the *embodied-ness* of the consciousnessbal kind.

The most amazing thing about both these kinds of *embodied-ness* of human being's consciousness is that they both look, appear, or seem as being *one hundred percent* genuine, authentic, real or, satya, or, as being *one hundred percent* actual, factual, material, substantial, objective, solid, or concrete, plus palpable, feelable, or, touchable, within the confines of their respective surrounding, milieu, context, backdrop, setting, locale, turfs, territory, arena, sphere, or domain.

To repeat.

The most amazing thing is that, both these kinds of *embodied-ness* of human being's consciousness, look, appear, or seem as being *one hundred percent* genuine, authentic, real or, satya, or, as being *one hundred percent* actual, factual, material, substantial, objective, solid, or concrete, plus palpable, feelable, or, touchable, within the confines of their respective surrounding, milieu, context, backdrop, setting, locale, turfs, territory, arena, sphere, or domain.

For example, the physical, material, substantial, or, concrete kind of *embodied-ness* of human being's consciousness looks, appears, or seems as being *one hundred percent* genuine, authentic, real or, satya, or, as being *one hundred percent* actual, factual, material, substantial, objective, solid, or concrete, plus palpable, feelable, or, touchable, during the wakeful state of human being's consciousness.

On the other hand, the consciousnessbal kind of *embodied-ness* belonging to the objectively *embodied,* twin, clone, incarnation or avatar of human being's consciousness during the latter's dream sleep state, looks, appears, or seems as being *one hundred percent* genuine, authentic, real or, satya, or, as being *one hundred percent* actual, factual, material, substantial, objective, solid, or concrete, plus palpable, feelable, or, touchable during the dream sleep state of human being's consciousness.

Does the above described, very stunning, startling, wonderful, or, wondrous truth, fact, or reality with regards to, or, vis-a-vis both kinds, types, or varieties of *embodied-ness* of human being's consciousness i.e., that both kinds, types, or varieties of *embodied-ness* of human being's consciousness look, seem or, appear objective in equal measure to two different forms, versions, or variants of the same, very same, selfsame, or, one and the same human consciousness, inside the confines or the boundaries of their respective surrounding, milieu, context, backdrop, setting, locale, turfs, territory, arena, sphere, or domain, not hint, suggest, convey, impart, or, point to the most fundamental fact about the *nature* of all the *phenomena, entities,* or, *things,* encompassed or hemmed in by the term or expression *"existence"*? This most fundamental fact about the *nature* of all the *phenomena, entities,* or, *things,* encompassed or hemmed in by the term or expression *"existence"*, is the following :-

"All phenomena, entities, things, or, existences, which are either merely observed, seen, visualised, or watched by human being's consciousness, or, are not merely observed, seen, visualised, or watched by human being's consciousness but are also palpated, felt, touched, tasted, smelled, or, heard, that is to say, perceived and experienced to the *fullest extent* by human being's consciousness, are *quintessentially, paradigmatically, archetypically, fundamentally,* or, *ultimately,* consciousnessbal and consciousnessbal in *nature* or *essence* and nothing else.

To paraphrase.

" All phenomena, entities, things, or, existences, which are either merely observed, seen, visualised, or watched by human being's consciousness, or, are not merely observed, seen, visualised, or watched by human being's consciousness but are also palpated, felt, touched, tasted, smelled, or, heard, that is to say, perceived and experienced to the *fullest extent* by human being's consciousness, are *quintessentially, paradigmatically, archetypically, fundamentally,* or, *ultimately* are consciousness and consciousness only and nothing else".

In other words, the physical, material, substantial, objective, or, concrete *cosmos, world,* or *universe* which human being's consciousness during its wakeful state, through the *agency, medium,* or *vehicle* of its *physically, materially, substantially,* or *concretely* embodied variant, form, or version, perforce, willy-nilly, without a choice, whether desired or not, or, every which way, is made or obliged to observe, see, visualise, watch, feel, touch, taste, hear and smell, as well as take part or participate in its wakeful state is nothing but consciousness and consciousness in *nature* or *essence,* albeit consciousness in its condensed, congealed, compressed, or, compacted *form* or *version.*

And this, one-of-a-kind, or, unique, consciousness is none other than the *ubiquitous* and the *infinite* field of consciousness called cosmic space, which in turn, is nothing but the expanded, distended, dilated, or inflated form or version of the dimensionless form or version or, of the original, earliest, starting, beginning, initial, first, primary, primeval, or primordial, form or version of God or Brahman aka the Source or Spring of human being's consciousness in the cosmos, on one hand, and the Author, Maker, or Creator, or, the Progenitor, Primogenitor, or Begetter, Author, Maker, Creator, Begetter, Procreator, Progenitor, or, Primogenitor, (Creator or Maker) of the physical matter in the cosmos, on the other.

Now let one pick up the previous thread once again from where one left it earlier.

In the *phenomenon* or the *magic-show* called *"dream world", "fantasy world", "dreamal world"* or *"fantasmal world"* of the dream sleep state of human being's consciousness, the *subjectively* bodiless, *twin, clone, incarnation, or avatar* of human being's consciousness, perforce, willy-nilly, without a choice, whether desired or not, or, every which way, is made or obliged to merely observe, see, visualise, or watch, its consciousnessbally *embodied* or, if one prefers, its objectively *embodied* counterpart, opposite number, vis-a-vis, or similitude, i.e., its twin, clone, incarnation or avatar, in addition to observing, seeing, visualising, or watching, the rest of the items and activities, which obtain, occur, or, exist plus happen or take place, inside the same, very same, selfsame, or, one and the same phenomenon or *magic-*

show called *"dream world", "fantasy world"*, *"dreamal world"* or *"fantasmal world"* of the dream sleep state of human being's consciousness.

This feat or achievement of observing, seeing, visualising, or watching its objectively *embodied* counterpart, opposite number, vis-a-vis, or similitude viz., its objectively *embodied* twin, clone, incarnation or avatar, one one hand, and observing, seeing, visualising or watching the rest of the items, and activities, which obtain, occur or exist, plus happen or take place inside the same, very same, selfsame, or, one and the same phenomenon or the *magic-show* called *"dream world", "fantasy world"*, *"dreamal world"* or *"fantasmal world"* of the dream sleep state of human being's consciousness, on the other, the *subjectively* bodiless, *twin, clone, incarnation, or avatar* of human being's consciousness, achieves or accomplishes, with the aid or assistance of its *consciousnessbal* kind, variety, or type of observing, seeing, visualising, or watching ability, capacity, competence, skill or talent and not with the aid or assistance of any kind of physical, material, substantial or concrete *senses* or *sense organs,* as it does during its wakeful state.

The consciousnessbally *embodied* or, if one prefers, objectively *embodied* counterpart or correlate, i.e., twin, clone, incarnation or avatar of human being's consciousness during the latter's dream sleep state, perforce, willy-nilly, without a choice, whether desired or not, or, every which way, is made or obliged, not merely to observe, see, visualise, or watch, but is also, perforce, willy-nilly, without a choice, whether desired or not, or, every which way, made or obliged to feel, touch, taste, hear, and smell, that is to say, perceive and experience *to the fullest extent,* the same, very same, selfsame, or, one and the same *phenomenon* or *magic-show* called the *"dream world", "fantasy world"*, *"dreamal world",* or *"fantasmal world"* every night, during the dream sleep state of human being's consciousness, through the *aid or assistance* of all its *five,* objective-looking, but *consciousnessbally-made* or *consciousnessbal-substance-made, sense organs* of its objective-looking but *consciousnessbally-made* or *consciousnessbal-substance-made body.*

Furthermore, the consciousnessbally *embodied* or, if one prefers, objectively *embodied* twin, clone, incarnation or avatar of human being's consciousness, during latter's dream sleep state, perforce, willy-nilly, without a choice, whether desired or not, or, every which way, is also made or obliged to take part or participate *to the fullest extent,* in the same, very same, selfsame, or, one and the same *phenomenon* or *magic-show* called *"dream world", "fantasy world"*, *"dreamal world",* or *"fantasmal world"* every night, during the dream sleep state of human being's consciousness, through the *aid or assistance* of the same, very same, selfsame, or, one and the same, *five* objective-looking but *consciousnessbally-made* or *consciousnessbal-substance-made sense organs* of its same, very same, selfsame, or, one and the same objective-looking but *consciousnessbally-made* or *consciousnessbal-substance-made body.*

There is another encounter or run into, which human being's consciousness has to face, meet, confront, or cope, willy-nilly, perforce, without a choice, whether desired or not, or, every which way, each and every day.

Human being's consciousness faces, meets, or confronts this second encounter or run into, willy-nilly, perforce, without a choice, whether desired or not, or, every which way, each and every day, during its wakeful state, while it is in its physically, materially, or substantially *embodied,* form or version, or, variant or rendition.

To repeat.

This second encounter or run into, human being's consciousness faces, meets, or confronts willy-nilly, perforce, without a choice, whether desired or not, or, every which way, during its wakeful state each and every day, while it is in its physically, materially, or substantially, *embodied* form or version, or, variant or rendition.

During this second encounter or run into, the physically, materially, or substantially, *embodied* form, version, variant or rendition, of human being's consciousness, each and every day, during its wakeful state, is made or obliged, willy-nilly, perforce, without a choice, whether desired or not, or, every which way, to observe, see, visualise, watch, touch, taste, hear, and smell, that is to say, perceive and experience to the fullest extent, plus participate or take part fully, all out, or, in all respects, in another *phenomenon* or *magic-show* called "physical, material, substantial or, concrete cosmos, world, or, universe", which it constantly or all the time, inaccurately, incorrectly, illiterately, nesciently, mistakenly, or, wrongly labels or designates as one hundred percent "actual", "factual", "real", "genuine", "authentic" or, "satya", on account of its thoughtlessness, impulsiveness, or, rashness plus, on account of the lack of due deliberation and diligence on its part.

The reality is that the phenomenon or the *magic-show* which human being's consciousness labels or tags as *"physical, material, substantial,* or *concrete cosmos, world, or universe"* and as one hundred percent "actual", "factual", "real", "genuine", "authentic" or, "satya", on account of its thoughtlessness, impulsiveness, or, rashness plus, on account of the lack of due deliberation and diligence on its part, and which, it is made or obliged to observe, see, visualise, watch, feel, palpate, touch, taste, smell, and hear, that is to say, perceive and experience to the fullest extent, plus in which it is made or obliged to participate or take part to the fullest extent, willy-nilly, perforce, without a choice, whether desired or not, or, every which way, each and every day, while it is in its wakeful state, is also only "partially real", "to a certain degree real " or, "real only when it obtains, occurs, or exists but not otherwise" i.e., is ephemeral, evanescent, episodic, fleeting, passing, short-lived, transient, temporary or impermanent in the manner the phenomenon or the *magic-show* called *"dream world", "fantasy world, "dreamal world",* or *"phantasmal world"* of its dream sleep state is, which human being's consciousness correctly, justly, or rightly calls a mere *"dream", "dreamry", "imagery"* or *"fantasy",* and, correctly, justly, or rightly, designates or denominates as only "partially real", "to a certain degree real " or, "real only when it obtains, occurs, or exists but not otherwise", viz., and, correctly, justly, or rightly, interprets, views, defines, or describes as only ephemeral, evanescent, episodic, fleeting, passing, short-lived, transient, temporary or impermanent.

To repeat.

The phenomenon or the *magic-show* of its dream sleep

state, human being's consciousness, correctly, justly, or rightly, labels or tags as a mere *'dream", "dreamry", ""imagery"* or *"fantasy"* or, as a mere *"dream world", "dreamry world", ""imagery world", "fantasy world","dreamal world",* or *"fantasmal world"* and, correctly, justly, or rightly, designates or denominates as only "partially real", "to a certain degree real " or, "real only when it obtains, occurs, or exists but not otherwise", viz., and, correctly, justly, or rightly defines, describes, interprets, or views as only ephemeral, evanescent, episodic, fleeting, passing, short-lived, transient, temporary or impermanent.

However, human being's consciousness must retrieve, recall, recollect, or remember one extremely important fact at this juncture which is the following.

Human being's consciousness does not label or tag the *phenomenon* or the *magic-show* of its dream sleep state as a mere *"dream", "dreamry", ""imagery"* or *"fantasy"* or, as a mere *"dream world", "dreamry world", "imagery world", "fantasy world", ""dreamal world",* or *"fantasmal world",* and therefore, only "partially real", "to a certain degree real " or, "real only when it obtains, occurs, or exists but not otherwise", i.e., ephemeral, evanescent, episodic, fleeting, passing, short-lived, transient, temporary or impermanent while it is still observing, seeing, visualising, watching, feeling, palpating, touching, tasting, smelling, or hearing it, that is to say, perceiving and experiencing it plus participating in it, but only later on, following its waking up from its dream and following the application of due deliberation and due diligence on its part during its wakeful state.

What has been said above can be put in another way.

The reality is that the *phenomenon* or the *magic-show* which human being's consciousness during its wakeful state and in its *physically, materially,* or, *substantially* embodied form, version, variant, or, rendition, willy-nilly, perforce, without a choice, whether desired or not, or, every which way, observes, sees, visualises, watches, feels, palpates, touches, tastes, smells, and hears, that is to say, perceives and experiences to the fullest extent, each and every day, plus in which it participates or takes part to the fullest extent, each and every day, and which it incorrectly, illiterately, nesciently, mistakenly, or wrongly calls or addresses by the name of *"physical, material, substantial,* or *concrete"* cosmos, world, or universe and on which, it incorrectly, illiterately, nesciently, mistakenly, or wrongly affixes, attaches, or fastens the label or tag that it is one hundred percent "actual', "factual", "real", "genuine", "authentic" or "satya", is nothing of the sort. It too is merely a "mithya" That is to say, it too is merely "partially real", "to a certain degree real " or, "real only when it obtains, occurs, or exists but not otherwise". In other words, it too is merely ephemeral, evanescent, episodic, fleeting, passing, short-lived, transient, temporary or impermanent in the manner the other *phenomenon* or the *magic-show* is, namely, the *phenomenon* or the *magic-show* which it also encounters, meets, or faces plus observes, sees, visualises, watches, feels, palpates, touches, tastes, smells, and hears, that is to say, perceives and experiences to the fullest extent, each and every day, plus in which it participates or takes part to the fullest extent, each and every day, albeit during its dream sleep state, and albeit through the agency, medium, or, instrumentality of its two *twins, clones, incarnations,* or, *avatars,* namely, the bodiless *twin, clone, incarnation,* or, *avatar* on one hand, and the *consciousnessbally* embodied *twin, clone, incarnation,* or, *avatar,* on the other, willy-nilly, perforce, without a choice, whether desired or not, or, every which way.

To paraphrase.

Both *phenomena* or *magic shows,* which human being's consciousness, encounters, meets, or faces, willy-nilly, perforce, without a choice, whether desired or not, or, every which way, each and every day, plus observes, sees, visualises, watches, feels, palpates, touches, tastes, smells, and hears, that is to say, perceives and experiences to the fullest extent, willy-nilly, perforce, without a choice, whether desired or not, or, every which way, each and every day, plus in which it participates or takes part to the fullest extent willy-nilly, perforce, without a choice, whether desired or not, or, every which way, each and every day, namely, the *physical, material* or *substantial* cosmos, world, or, universe of human consciousness's wakeful state on one hand, and the *"dreamal"* or the *"phantasmal"* cosmos, world, or, universe of human consciousness's dream sleep state, on the other, are not "satya", that is to say, are not eternal, immortal, timeless, deathless, or, endless, in the manner Cosmic Space, God, or, Brahman is, viz., in the manner, the Supreme Source or Spring of human consciousness on one hand, and the Supreme Creator or Maker of the *physical matter,* on the other hand, is, or, better still, in the manner, the Supreme Source or Spring as well as the Supreme Author, Maker, or Creator, or, the Progenitor, Primogenitor, or Begetter, Creator or Maker of the two phenomena or magic shows in question is, or, in the manner, the Supreme Source or Spring as well as the Supreme Author, Maker, or Creator, or, the Progenitor, Primogenitor, or Begetter, Author, Maker, Creator, Begetter, Procreator, Progenitor, or, Primogenitor, (Creator or Maker) of the two phenomena or magic shows being discussed is.

There are three Sanskrit *terms, expressions,* or, *words,* which are often used in the discipline, domain, or, arena, of Adwait-Vedanta. These are :-

Satya.
Mithya.
Asatya.

These three Sanskrit *words, terms,* or *expressions* have very precise meaning and therefore, they must not be used loosely i.e., incorrectly.

Satya is defined as that which is eternal, immortal, timeless, deathless, or, endless. There is only one *Satya,* namely, Cosmic Space, God, or, Brahman viz., the Supreme Source or Spring of human consciousness, on one hand, and, the Supreme Author, Maker, or Creator, or, the Progenitor, Primogenitor, or Begetter, Author, Maker, Creator, Begetter, Procreator, Progenitor, or, Primogenitor, (Creator or Maker) of the *physical matter,* on the other, or, better still, the Supreme Source or Spring, as well as, the Supreme Author, Maker, or Creator, or, the Progenitor, Primogenitor, or Begetter, Author, Maker, Creator, Begetter, Procreator, Progenitor, or, Primogenitor, (Creator or Maker), of the two phenomena or magic shows in question or, the Supreme Source or Spring, as well as, the Supreme Author, Maker, or Creator, or, the Progenitor, Primogenitor, or Begetter, Author, Maker, Creator, Begetter, Procreator, Progenitor, or, Primogenitor, (Creator or Maker), of the two phenomena or magic shows, being discussed.

Mithya is that which is only "partially real", "to a certain degree real " or, "real only when it obtains, occurs, or exists but not otherwise". In other words, m*ithya* is that which is ephemeral, evanescent, episodic, fleeting, passing, short-lived, transient, temporary or impermanent. Both *phenomena* or *magic shows,* which human being's consciousness, encounters, meets, or faces, willy-nilly, perforce, without a choice, whether desired or not, or, every which way, each and every day, plus observes, sees, visualises, watches, feels, palpates, touches, tastes, smells, and hears, that is to say, perceives and experiences to the fullest extent, willy-nilly, perforce, without a choice, whether desired or not, or, every which way, each and every day, plus in which it participates or takes part to the fullest extent willy-nilly, perforce, without a choice, whether desired or not, or, every which way, each and every day, namely, the *physical, material* or *substantial* cosmos, world, or, universe of human consciousness's wakeful state on one hand, and the *"dreamal"* or the *"phantasmal"* cosmos, world, or, universe of human consciousness's dream sleep state, on the other are "Mithya" and nothing else. As said before, these two *phenomena* or *magic shows* are not "Satya", that is to say, are not eternal, immortal, timeless, deathless, or, endless, in the manner Cosmic Space, God, or, Brahman is, viz., in the manner, the Supreme Source or Spring of human being's consciousness and the Supreme Author, Maker, or Creator, or, the Progenitor, Primogenitor, or Begetter, Author, Maker, Creator, Begetter, Procreator, Progenitor, or, Primogenitor, (Creator or Maker) of the *physical matter* is, or, better still, in the manner, the Supreme Source or Spring as well as the Supreme Author, Maker, or Creator, or, the Progenitor, Primogenitor, or Begetter, (Creator or Maker) of the two phenomena or magic shows in question is or, in the manner, the Supreme Source or Spring as well as the Supreme Author, Maker, or Creator, or, the Progenitor, Primogenitor, or Begetter, Author, Maker, Creator, Begetter, Procreator, Progenitor, or, Primogenitor, (Creator or Maker) of the two phenomena or magic shows being discussed is.

Asatya means "absolute untruth" or "absolute lie" or, that which never existed.

The phenomenon or *magic-show* which human being's consciousness encounters, meets, or faces, willy-nilly, perforce, without a choice, whether desired or not, or, every which way, during its dream sleep state, each and every night, is also observed, seen, visualised, watched, touched, tasted, heard and smelled by it, that is to say, perceived and experienced to the fullest extent by it, plus participated or taken part to the fullest extent by it, during its dream sleep state, each and every night, willy-nilly, perforce, without a choice, whether desired or not, or, every which way, through the medium, means, or instrumentality of its two *twins, clones, incarnations, or avatars,* namely its *subjectively* bodiless, *twin, clone, incarnation, or avatar* on one hand, and its *objectively* embodied viz., *consciousnessbally* embodied *twin, clone, incarnation, or avatar,* on the other, during its dream sleep state and which it rightly, correctly, or justly calls or labels as being a mere "dream", "dreamry", "imagery", or "fantasy" or, *"dream world", "fantasy world", "dreamal world",* or *"fantasmal world"* and, which it rightly, correctly, or justly, designates, describes, or addresses as being only "partially real", "to a certain degree real"or, "real only when it obtains, occurs, or exists but not otherwise" i.e., as being ephemeral, evanescent, episodic, fleeting, passing, short-lived, transient, temporary or impermanent, not while it is engaged in observing, seeing, visualising, watching, tasting, touching, hearing and smelling it, that is to say, perceiving and experiencing it to the fullest extent, plus, engaged in participating or taking part in it, to the fullest extent, but only later on, that is to say, but only on waking up from its "dream" i.e., following its entry or ingress into its wakeful state and application of due deliberation and due diligence on its part vis-a-vis or with regards to the phenomenon or the magic show in question.

Therefore, it is incumbent or, necessary as a duty on the

part of human being's consciousness during its wakeful state, when it is in its *physically, materially,* or, *substantially,* embodied *form, version, incarnation,* or *avatar,* to employ due deliberation and due diligence, vis-a-vis or, with regard to, the *phenomenon* or *magic-show* which it willy-nilly, perforce, without a choice, whether desired or not, or, every which way, each and every day during its wakeful state, encounters, meets, or faces, plus, willy-nilly, perforce, without a choice, whether desired or not, or, every which way, each and every day, observes, sees, visualises, watches, feels, palpates, touches, tastes, hears, and smells, that is to say, perceives, and experiences to the fullest extent, and in which it participates or, takes part to the fullest extent, willy-nilly, perforce, without a choice, whether desired or not, or, every which way, each and every day, and, on which it inaccurately, incorrectly, illiterately, nesciently, mistakenly, or wrongly, affixes, attaches, or fastens the label or tag, that, it is *"physical, material, substantial,* or *concrete"* in nature, on one hand, and it is one hundred percent "actual", "factual", "real", "genuine", "authentic", or, "Satya", on the other, on account of its thoughtlessness, impulsiveness, or, rashness plus, on account of the lack of due deliberation and due diligence on its part, or, on account of the lack of due assiduousness, conscientiousness, carefulness, meticulousness, or thoroughness, on its part.

If human being's consciousness during its wakeful state, cares to apply due assiduousness, conscientiousness, carefulness, meticulousness, or thoroughness with regard to finding out the true nature of the *phenomenon* or *magic-show* which it willy-nilly, perforce, without a choice, whether desired or not, or, every which way, each and every day, during its wakeful state, encounters, meets, or comes in contact with, plus, observes, sees, visualises, watches, feels, palpates, touches, tastes, hears, and smells, that is to say, perceives, and experiences to the fullest extent, and in which it participates or, takes part to the fullest extent, each and every day, and, on which it inaccurately, incorrectly, illiterately, mistakenly, or wrongly, affixes, attaches, or fastens the label or tag, that, it is *"physical, material, substantial,* or *concrete"* in nature, on one hand, and that it is one hundred percent "actual", "factual", "real", "genuine", "authentic", or, "Satya", on the other, it will one day discover to its great surprise that it is nothing of the sort. That is to say, it quintessentially is neither *"physical, material, substantial,* or *concrete"* in nature, nor one hundred percent "actual", "factual", "real", "genuine", "authentic", or, "Satya". Instead, it quintessentially is "*consciousnessbal, awarenessbal,* sentiencel, dreamal, or *phantasmal"* in nature because the *physical matter* aspect of its constitution, composition, or make-up is made or composed of condensed, congealed, compressed, or compacted form or version of a segment, section, part, or portion of the *ubiquitous and infinite* field of consciousness aka Cosmic Space aka God or Brahman aka the Source or Spring of human being's consciousness, on one hand, and, the Author, Maker, or Creator, or, the Progenitor, Primogenitor, or Begetter, Author, Maker, Creator, Begetter, Procreator, Progenitor, or, Primogenitor, (Creator or Maker) of physical matter, on the other. This is the fact, because this phenomenon or magic show in question, or, under consideration or discussion, is merely, purely, or simply a day-dream or oneiric of the *ubiquitous and infinite* field of consciousness aka Cosmic Space aka God or Brahman aka the Source or Spring human being's consciousness, on one hand, and, the Author, Maker, or Creator, or, the Progenitor, Primogenitor, or Begetter, Author, Maker, Creator, Begetter, Procreator, Progenitor, or, Primogenitor, (Creator or Maker) of physical matter, on the other, nothing more nothing less.

The accomplishment or achievement of the correct kind of conclusion, judgement, or decision, vis-a-vis or, with regards to the true nature of the *phenomenon* or *magic-show* in question, under discussion or, under consideration, which human being's consciousness, in its *physically, materially,* or, *substantially* embodied *form, version, incarnation,* or *avatar,* willy-nilly, perforce, without a choice, whether desired or not, or, every which way, each and every day, observes, sees, visualises, watches, feels, palpates, touches, tastes, hears, and smells, that is to say, perceives, and experiences to the fullest extent, and, in which it participates or takes part to the fullest extent each and every day during its wakeful state, plus on which it inaccurately, incorrectly, illiterately, mistakenly, or wrongly affixes, attaches, or fastens the label or tag that it is *"physical, material, substantial,* or *concrete"* in nature, on one hand, and "actual", "factual", "real", "genuine", "authentic", or "satya", on the other, due to its thoughtlessness, impulsiveness, or, rashness plus, on account of the lack of due deliberation and due diligence on its part, or, on account of the lack of due assiduousness, conscientiousness, carefulness, meticulousness, or thoroughness on its part, is very important if human being's consciousness in its embodied *form, version, incarnation,* or *avatar,* wants to know or learn about Cosmic Space or Brahmandic Aakash i.e., about God or Brahman, who is the Absolute, Ultimate, or, Supreme Truth inside the *cosmos* as well as beyond the *cosmos* plus who is timeless, endless, perpetual or eternal, and who is the *Source* or *Spring* of human being's consciousness in the cosmos, on one hand, and the Author, Maker, or Creator, or, the Progenitor, Primogenitor, or Begetter, *Creator* or *Maker* of the *physical matter* in the cosmos, on the other, that is to say, who is the *Source* or *Spring* as well as the Author, Maker, or Creator, or, the Progenitor, Primogenitor, or Begetter,

Author, Maker, Creator, Begetter, Procreator, Progenitor, or, Primogenitor, (*Creator* or *Maker)* of the phenomenon or magic show in question, under discussion, or, under consideration, i.e., the phenomenon or *magic-show* which human being's consciousness in its embodied *form, version, incarnation,* or *avatar,* during its wakeful state, willy-nilly, perforce, without a choice, whether desired or not, or, every which way, each and every day, observes, sees, visualises, watches, feels, palpates, touches, tastes, hears, and smells, that is to say, perceives, and experiences to the fullest extent, and in which it participates or takes part to the fullest extent, each and every day, during its wakeful state, and, on which it inaccurately, incorrectly, illiterately, mistakenly, or wrongly affixes, attaches, or fastens the label or tag that it is *"physical, material, substantial,* or *concrete"* in nature, on one hand, and, "actual", "factual", "real", "genuine", "authentic", or "satya", on the other, due to its thoughtlessness, impulsiveness, or, rashness plus, on account of the lack of due deliberation and due diligence on its part, or, on account of the lack of due assiduousness, conscientiousness, carefulness, meticulousness, or thoroughness on its part.

The *four expressions* employed above, namely Cosmic Space, Brahmandic Aakash, God and Brahman, allude to, convey, connote, denote, express, indicate, imply, refer to, spell out, or, stand for the *same, very same, selfsame,* or, *one and the same,* Absolute, Ultimate, or Supreme Truth which is *eternal, immortal, undying, unfading, deathless, or amaranthine.* That is to say, which is not *transient, temporary, dying or decaying* in the way or manner the above narrated two phenomena or magic shows are, namely the one which human being's consciousness, willy-nilly, perforce, without a choice, whether desired or not, or, every which way, each and every day during its dream sleep state, observes, sees, visualises, watches, feels, palpates, touches, tastes, hears, and smells, that is to say, perceives, and experiences to the fullest extent, and, in which it participates or takes part to the fullest extent, each and every day, during its dream sleep state, and which it correctly, justly, or, rightly, calls or tags as *"dream"* and which it correctly, justly, or, rightly, epithets or labels as that which is only or merely, *"partially real", "to a certain degree real " or, "real only when it obtains, occurs, or exists but not otherwise"* i.e., which is awfully, dreadfully, or, frightfully, ephemeral, evanescent, episodic, fleeting, passing, short-lived, transient, temporary or impermanent, not while it observes, sees, visualises, watches, perceives, experiences, and participates or takes part in it, but later on, following due deliberation and due diligence on its part, and, the other one, or, the *other phenomenon* or *magic-show,* which human being's consciousness, each and every day, during its wakeful state, willy-nilly, perforce, without a choice, whether desired or not, or, every which way, observes, sees, visualises, watches, perceives, experiences and in which it participates or takes part, each and every day, during its wakeful state, and which it wrongly, illiterately, nesciently, or ignorantly calls or tags as *"physical, material, substantial* or *concrete"* in nature, and, which it wrongly, illiterately, nesciently, or ignorantly, epithets or labels as "actual", "factual", "real", "genuine", "authentic", or, "satya", on account of its thoughtlessness, impulsiveness, or, rashness, plus, on account of the lack of due deliberation and due diligence on its part or, on account of the lack of due assiduousness, carefulness, conscientiousness, meticulousness, or thoroughness on its part.

The *creator,* or, if one prefers, the source of both the above described phenomena or magic shows i.e., the phenomenon or magic show belonging to the dream sleep state of human being's consciousness on one hand, and the phenomenon or magic show belonging to the wakeful state of human being's consciousness on the other, is the Same, Very Same, Self-Same, or, One and The Same, Conscious Being and not two separate or different conscious beings as universally believed by human consciousnesses.

And this One and Only, Unique, Singular, Matchless, or, Unparalleled Being is the Absolute, Ultimate, or

Supreme Being which is *eternal, immortal, undying, unfading, deathless, or amaranthine.* That is to say, *IT* is not *transient, temporary, dying or decaying* in the way or manner the above narrated two phenomena or magic shows are.

This O*ne and Only,* Absolute, Ultimate, or Supreme Being, which is *eternal, immortal, deathless, amaranthine, undying, or unfading,* is called Cosmic Space, Brahmandic Aakash, God, or Brahman. All these *four expressions,* as said before, allude to, convey, connote, denote, express, indicate, imply, refer to, spell out, or, stand for the *same*, *very same, selfsame,* or, *one and the same,* Absolute, Ultimate, or Supreme Truth which is *eternal, immortal, undying, unfading, deathless, or amaranthine.* That is to say, which is not *transient, temporary, dying or decaying* in the way or manner the above narrated two phenomena or magic shows are.

Cosmic Space, Brahmandic Aakash, God, or, Brahman, has made, composed, constructed, created, fashioned or forged, the brain of human beings in such a unique way and of such an extraordinary form, shape, format or design that it is able to retain the highest amount of *consciousness* which Cosmic Space, Brahmandic Aakash, God, or Brahman, grants or gifts to human being's brain, as compared to its other conscious, aware, or sentient scions, offsprings, or progenies.

As a result or consequence of Cosmic Space's, Brahmandic Aakash's, God's, or, Brahman's grant or gift of the highest amount of *consciousness* to human brain, as compared to its other conscious, aware, or sentient scions, offsprings, or, progenies, Homo sapiens, humankind, humanity, human race, or, human species has become the most *intelligent, insightful, intuitive, brilliant, clever, canny, gifted, perceptive, quick-minded, quick-witted, sharp, smart,* or *talented.*

This extremely high *intelligence, insightfulness, intuitiveness, brilliance, cleverness, canniness, giftedness, perceptiveness, quick-mindedness, quick-wittedness, sharpness, smartness,* or *talent* of human being's consciousness, awareness, sentience, or, mind, has made it the most inquisitive, inquiring, investigative, searching, questioning, poking, peering, nosy, or, analytical, amongst all the other conscious, aware, sentient, or, mindful beings of Cosmic Space's, Brahmandic Aakash's, God's, or Brahman's *creation.* This inquisitiveness, curiousness, or, nosiness of human being's consciousness, prompts, makes, or, encourages, it ask such extremely difficult questions as the following :-

"What is the true nature of the physical world?"

"Who or what is the source or spring or, who or what is

the creator or maker of the physical world?"

"How was this physical world made or created?"

"Why was this physical world made or created?"

"When was this physical world made or created?"

"How, why, and when will this physical world die or, come to an end?"

"What is the nature of cosmic space?"

"What is the nature of human consciousness?"

"Who or what is the source or spring of cosmic space?"

"Who or what is the source or spring of human consciousness?"

Why did cosmic space come into being in the physical cosmos?"

"Why did human consciousness come into being in the physical cosmos?"

"Will human consciousness disappear from the physical cosmos one day or, any time in the future?"

"If human consciousness will disappear from the physical cosmos one day or, any time in the future, where will it go or, where will it abide or reside?" Or, will it die one day in the manner the human body is destined to die one day.

"What happens to human beings after death?"

"What is the actual nature of the human death?"

And more.

Cosmic space, Brahmandic Aakash, God, or, Brahman, knew fully well that human beings will ask all the above, extremely difficult questions and much more, on account of the huge amount of consciousness It has granted or gifted them via their brain, and, due to the consequent possession on human being's part, of an *extremely high degree* or *level* of *intelligence, insightfulness, intuitiveness, brilliance, cleverness, canniness, giftedness, perceptiveness, quick-mindedness, quick-wittedness, sharpness, smartness* or *talent.*

Cosmic space, Brahmandic Aakash, God, or, Brahman, therefore, introduced into the lives of human beings, the daily experience of "*dreams", "dreamries", "imageries, or "fantasies"* or, the daily experience of *"dreamal or phantasmal cosmoses worlds, or universes"* during their sleep.

Or, absolutely to the point, Cosmic space, Brahmandic Aakash, God, or, Brahman, therefore, introduced into the lives of human beings the daily experience of dream sleep state, during their absolutely necessary or, vital *activity* of sleep each night, particularly their absolutely necessary or, vital *activity* of deep sleep each night, on one hand, plus the possession on their part, of the art, skill, ability, or, competence to daydream, reverie, or, oneiric at will, ad-lib, ad-libitum, without restraint, unrehearsed, at their pleasure or, according to their pleasure, or, as they wish, or, as they think best, during their wakeful state, on the other.

It is now up to human beings to make full use of Cosmic Space's, Brahmandic Aakash's, God's, or, Brahman's granted or gifted, two very special, extraordinary, or, exceptional boons, blessings, grace, or favours in order to arrive at the correct answers with regards to all the above extremely difficult questions which they constantly pose to themselves throughout their lives.

ABOUT THE AUTHOR

Dr. Chandra Bhan Gupta, was born and educated in Lucknow, India.

He commenced his medical career in India with several notable medical articles to his credit.

Subsequently, he went to UK., where he continued his distinguished medical career, gaining the highest postgraduate and honorary accolades within his field.

Such questions as how and why man and the rest of creation have come into being, as well as the true nature of the creator and where is his abode, troubled him from an early age.

In an attempt to find answers to these eternal questions, he went through extreme austerities or penance over the course of many years, accompanied by long periods of deep meditation.

Enlightenment from the Almighty came in 1995, which resulted in the writing of the first book on the theme of "Supra-Spirituality", called 'Adwaita Rahasya: Secrets of Creation Revealed', followed by two more books delving deeper into the same theme, entitled 'Space is The Mind of God: A Scientific Explanation of God and His Abode', and the present series of works 'Cosmic Space is God and Universe is God's Dream'.

Printed in Poland
by Amazon Fulfillment
Poland Sp. z o.o., Wrocław

54779742R00204